EGYPTIAN HIERATIC TEXTS

TRANSCRIBED, TRANSLATED AND ANNOTATED

BY

ALAN H. GARDINER D. LITT.
LAYCOCK STUDENT OF EGYPTOLOGY AT WORCESTER COLLEGE, OXFORD

SERIES I: LITERARY TEXTS OF THE NEW KINGDOM

PART I

THE PAPYRUS ANASTASI I AND THE PAPYRUS KOLLER, TOGETHER WITH THE PARALLEL TEXTS

PROVISIONAL PREFATORY NOTICE

The following extracts from the Prospectus in which the present work is announced will explain its scope and arrangement:—

"It is proposed to divide the whole material into a number of series, each dealing with a different class of text: the first series will contain *Literary Texts*, the subject of the second will be *Magical Texts*; other series will be devoted to *Business Documents*, *Juristic Texts*, *Letters*, etc. Every series will comprise a certain number of parts, each of which will be complete in itself, containing one or more entire texts, together with bibliographical introductions, translations and notes. When a sufficient number of parts in any series has been issued, a volume will be concluded; the separate volumes will include full indices to the words discussed in the notes, and to personal, royal and geographical names; and the numbering of the pages will be so arranged that the printed portion containing the translations and introductions, and the autographed portion containing the text, may be bound up separately.

"In the autographed portion, the left-hand page will contain the hieroglyphic transscript, while the palaeographical comments will occupy the opposite page. The direction and mutual position of the signs in the original will be retained in the transcript, in order to facilitate comparison with the hieratic. For economy of space and for other reasons the text will not be split up into sentences; nor will the restoration of lacunae be undertaken beyond a limited extent. Where more than one manuscript is preserved, the parallel texts will be reproduced *in extenso* side by side. It need hardly be said that the series aims at the most exemplary accuracy, particularly in the establishment of the text, the originals having been diligently collated in almost every case.

"The printed portion will consist of bibliographical introductions, complete English translations, and such philological notes, with abundant references, as can be given without the use of hieroglyphic type. In the preparation of the translations all previous renderings will be carefully compared, so that no earlier suggestions of value may be ignored. The translations will err rather on the side of literalness than on that of freedom, but the notes will supply a certain corrective in the way of paraphrases and explanations of passages where the English rendering is obscure."

Title-pages, Preface and Indices will follow later at the conclusion of Volume I.

For the immediate convenience of the reader the following facts should be noted. The printed portion of the work is separately paged with numbers accompanied by asterisks (1*, 2*, 3*), and in seeking cross-references in the foot-notes this fact is strictly to be borne in mind. Such references as p. 5, line 9, or p. 5a, note 11d refer to the autographed text and to the autographed notes accompanying the same; and the page-numbers thus alluded to are those that will be found following the words *Lit. Texts* at the top right-hand corner of the autographed pages; the numbers at the bottom of the autographed pages will be omitted in subsequent parts, and should be disregarded altogether. The double mode of pagination has been adopted in order that the printed and autographed portions may later be bound up separately.

The references to Egyptological literature are given in the forms commonly used in Egyptological books. The Semitic scholar should note that MAX MÜLLER, *As. u. Eur.* = W. MAX MÜLLER, *Asien und Europa nach Altägyptischen Denkmälern*, Leipzig 1893; and that BURCH. stands for MAX BURCHARDT, *Die altkanaanäischen Fremdworte und Eigennamen im Aegyptischen: zweiter (Schluss-) Teil, Listen der syllabisch geschriebenen Worte*, Leipzig 1910.

For the the transcription of Egyptian words the symbols used in the grammar of A. ERMAN have been retained, except [illegible]

I. A satirical letter.

Sources of the text. 1. The Papyrus Anastasi I (*Pap. Brit. Mus. 10247*) was purchased for the British Museum in 1839 from Signor ANASTASI, the Swedish Consul in Egypt. It measures 8 1/4 metres in length; its height is 20·5 cm., i. e. it is of the normal height of literary papyri of the second Theban period. Both in respect of size and of calligraphy the papyrus is one of the finest specimens of a Ramesside book. The text, written upon the horizontal fibres, comprises 28 pages of 8 or 9 lines apiece; at the beginning there is a protecting strip of 10·5 cm. The *verso* is uninscribed, save for a few large illegible signs at the back of the 19th. page. The papyrus contains but a single composition, which is complete except for the losses caused by lacunae; these are very abundant, especially in the earlier pages. The composition is divided into nineteen paragraphs, of which the first words are always written in red. Verse-points occur on the 9th. and 10th. pages. Corrections are to be found here and there; one written at the top of page 3 is probably due to a teacher's hand. The type of handwriting is closely similar to, but not identical with, that of Ennene, the scribe to whose industry we owe several of the finest of the London hieratic papyri. The papyrus *Anastasi I* may therefore be dated with approximate accuracy to the reign of Sethos II (see MÖLLER, *Paläographie* II 11)[1]. — A good hand-facsimile by J. NETHERCLIFT is published in the *Select Papyri in the hieratic character from the collections of the British Museum, London* 1842, plates XXXV—LXII[2]. In CHABAS' book *Le Voyage d'un Egyptien* the plates, comprising *Anast. I* 8; 18,3—8; 19 to 28, seem to have been merely reproduced from the official publication. — The transcript here given is based on that made by K. SETHE for the purposes of the Berlin hieroglyphic dictionary; this has been diligently collated with the original on several separate occasions[3].

2. Ostracon belonging to Professor Petrie (abbreviated O. P.). A slab of limestone 21×26 cm., containing a half-obliterated duplicate of *Anast. I* 1, 1—4 with some additions. Unpublished.

3. Ostracon in the Bibliothèque Nationale, Paris (abbrev. O. B. N.), a limestone tablet containing *Anast. I* 1,6—2,6 with numerous variants and additions. Published by W. SPIEGELBERG, *Beiträge zur Erklärung des Papyrus Anastasi I* in the *Zeitschrift für ägyptische Sprache* 44 (1908), 119—122 and Tafel V. VI, with (1) a photograph of the stone in its present much deteriorated condition, (2) a hand-copy of the hieratic made in 1892, and (3) a hieroglyphic transcription. My text is based on a comparison of these materials.

4. Ostracon in the Brussels collection (abbrev. O. Br.), a broken potsherd containing portions of *Anast. I* 2, 1—5 and closely following the ostracon of the Bibliothèque

1) According to MÖLLER Ennene was a scribe of the Memphite school. The same would doubtless hold good of the writer of *Anastasi I*.

2) This is sometimes referred to in my notes upon the text as "the facsimile".

3) The proofs of this edition were finally compared with the original in September 1910.

logical science. Here the first *Anastasi* papyrus obtained its due in the shape of an admirable analysis interspersed with translations (p. 508—513); and the real subject of the book was lucidly and convincingly proved to be a literary controversy between two scribes, the form adopted being that of a letter pretending to be the answer to an ill-worded and pretentious communication. ERMAN's general interpretation requires hardly any modification today, and his translations need but little change, though naturally the latter embrace only the more easily intelligible and picturesque passages.

For the past five and twenty years no new edition or translation has to be noted, so that for complete, or nearly complete, renderings of the papyrus recourse must still be had to the works of CHABAS and LAUTH. A hieroglyphic transcription of *Anast. I* 18,3 to 28,8 (the voyage) is given in E. A. W. BUDGE's *Egyptian Reading Book* (1888) p. 157—169, and in the larger edition of the same work (1896) p. 274—294. The geographical problems are minutely discussed in W. MAX MÜLLER's *Asien und Europa nach altägyptischen Denkmälern* (1893), particularly pp. 172—175. MASPERO has dealt with two passages of the voyage in his articles *Entre Joppé et Mageddo* in the *Études .... dédiées à M. le Dr. C. Leemans*, p. 4—6; and *Notes sur la géographie égyptienne de la Syrie, III* in the *Recueil de Travaux* 19, 68—73; SPIEGELBERG has given a rendering of, and commentary on, 9,4—10,6 under the title *Eine Probe der ägyptischen persönlichen Satire* in his *Beiträge zur Erklärung des Papyrus Anastasi I* (*Ä. Z.* 44 [1908], 118—125). A fine ostracon belonging to the Bibliothèque Nationale, Paris, and supplementing the beginning of the papyrus is also published in the last-named article.

General remarks. The composition contained in the first *Anastasi* papyrus was written certainly not earlier, but probably also not later, than the reign of Rameses II, the name of that king occurring in several passages (12,3; 18,8; 27,3. 5). Of the ten ostraca and papyri preserving greater or less portions of the text none is of more recent date than about the middle of the 20th. Dynasty; and this quite unusual number of manuscripts, scattered over so brief a period, bears eloquent testimony to the popularity which the work enjoyed in the Ramesside schools. Nor is its popularity hard to explain, if the standards of taste current in those times are carefully borne in mind. In the first place the theme upon which the entire composition turns is the profession of the scribe, and no lesson was more assiduously instilled into the mind of the Egyptian schoolboy than the belief in the dignity and the advantages of that career. Thus from one aspect *Anastasi I* ought to be regarded as akin to the numerous effusions in which the student is bidden apply himself diligently to the art of writing[1], or where other occupations are invidiously compared with the labours of the scribe[2]. Secondly, its wealth of topics and consequent variety of vocabulary must have given the text particular value as a model of style and as a means of teaching orthography. The abundant use made of foreign words and the display of erudition with regard to outlandish place-names agree well with what we know of the predilections of the age[3]. Lastly, the good-humoured raillery which is the dominant note of the papyrus springs from one of most attractive sides of the Egyptian temperament. Hints of the Egyptian's love of repartee and appreciation of irony may be found in the snatches of conversation written above the scenes on the walls of tombs, or in the paintings and sculptures themselves, or in the rare caricatures that have survived, and samples are to be found here and there in the literature[4]; but nowhere are these attributes more strikingly illustrated than in

---

1) See ERMAN, *Aegypten* p. 442 foll.
2) E. g. *Anast. IV* 9, 4 foll.
3) See ERMAN, *Aegypten* p. 683.
4) See especially the letter *Pap. Bibl. Nat.* 198, 2 published by SPIEGELBERG, *Correspondances du Temps des Rois-Prêtres* p. 68—74.

Anastasi I. If it must be confessed that the quality of the wit is poor, and that the satirical vein is intolerably insistent, still, that a Ramesside author should so well have understood to use language in a way not immediately suggested by its plain face-value is an achievement to be respected.

The whole character of the book would make it impossible to translate satisfactorily into any modern language, even if its vocabulary were more familiar to us. As it is, our ignorance of many of the actual words often leaves the sense totally obscure; and the difficulties are further increased by the frequency of the lacunae and by the textual corruptions. In the rare cases where more than two manuscripts are preserved *Anastasi I* is as a rule found to stand alone in its readings; the Brussels ostracon agrees closely with that of the Bibliothèque Nationale, the Cailliaud ostracon still more closely with that of the Louvre. Possibly the reason may be that *Anastasi I* is a Memphitic text[1], whereas the ostraca without exception are of Theban provenance. Perhaps of all the sources the Louvre ostracon has the least corrupt text; the most corrupt is certainly the Turin papyrus. As for *Anastasi I,* in many passages it is obviously in error (see for a particularly glaring case 4,3); but it would not be wise in the present state of our knowledge to pronounce a very definite verdict on the degree of its inaccuracy. The language is similar to that of other literary texts of the period, i. e. a mixture between the vulgar spoken dialect and the old classical style.

The argument. (I) The scribe Hori, a man of great erudition and celebrity, employed in the Royal stables, (II) writes to his friend the scribe Amenemope, (III) wishing him all prosperity in this world and all happiness in the next. (IV) Hori writes to say that his friend's letter reached him at a moment of leisure; his joy at its reception was lessened by a perusal of the contents, which appeared to him to be very mediocre. (V) In spite of his having called many helpers to his aid, Amenemope has contrived to make a bad job of the commission he was called upon to perform. (VI) Hori will now reply in a letter of the same kind, and will excel Amenemope at every point, dealing with the very same topics. At the outset Amenemope has had the discourtesy to omit the usual salutations. (VII) He has also expressed his contempt of Hori's ability. The latter replies ironically, naming various persons who have attained to positions of ease and wealth in spite of some ludicrous moral or physical defect; the exact point of the answer is not clear. (VIII) Hori goes on to poke fun at his friend's pretensions to learning and ends by advising him not to meddle with ancient lore. (IX) Accused of having no right to his title of scribe, Hori refers Amenemope to the official registers. (X) An obscure competition between the rivals. (XI) Again accused of being no true scribe, Hori asks that the god Onuris shall be appealed to as arbiter. (XII) When commanded to undertake any difficult calculation, Amenemope either turns to others for advice, or else fails to give any answer; this is illustrated by several examples, the building of a ramp, (XIII) the transport of an obelisk, (XIV) the erection of a colossus, and (XV) the furnishing of supplies for the troops on a foreign military expedition. (XVI) Amenemope has laid claim to the title of Maher; Hori replies by enumerating a number of places in Northern Syria never visited by Amenemope, and an imaginary picture of the latter enduring the discomforts of the Maher's life is conjured up. (XVII) Amenemope is questioned as to the towns of Phoenicia (XVIII) and as to the towns further southwards. (XIX) Sundry other places frequented by the Maher are named, and Amenemope is imagined as experiencing the adventures of the Maher's life — the traversing of a precipitous region, the danger from wild beasts, the breaking of the chariot, the arrival at Joppa, and finally the repairing of the chariot and the start on a new journey. (XX) Hori cross-examines his friend as to the places on the high-road to Gaza, and finds him totally ignorant of them.

1) See above p. 1*, footnote 1.

He regards his own superiority as now fully proved, and bids Amenemope not to be angry but to listen coolly; in this way he too will learn to discourse about foreign parts and the incidents of travel.

I. The rank and qualifications of the writer.

The scribe of noble parts, patient in discussion[1], at whose utterances men rejoice when they are heard, skilled in the hieroglyphs; there is nothing he does not know. He is a champion in valour and in the art of Seshyt[2]; servant of the lord of Khmūn[3] in the[4] hall of writing; assistant-teacher[5] in the office of writing[6]. First of his companions, foremost of his fellows, prince of his contemporaries, without his peer. His merit is proved[7] in every stripling. His hand goes forth (??)[8], his fingers magnify the young (??). Noble, keen of wit[9], adept 1,5 in knowledge; fortunate because of it (??)[10]. Protecting himself by his good qualities; beloved in (men's) hearts, without being repelled (?)[11]; men like to make a friend of him, they do not tire of him. Swift to inscribe[12] empty rolls. Youthful, eminent of charm, pleasant of grace[13]. Unravelling the obscurities of the annals like him who made them[14]. All that goes forth from his mouth is steeped in honey[15]; the heart is refreshed with it as it were with medicine. Groom[16] of his 2,1 Majesty, follower of the Sovereign, trainer of the steeds[17] of the king, ardent cultivator[18] of the stable[19]. The old man who doeth like him[20] is beaten (?)[21]. He

1) *Nḏnḏ-ri* only here; perhaps read *nḏ-ri*, as the *Petrie Ostracon* (*O. P.*) may have had; cf. too (*Lit. Texts*) p. 4, l. 6. 7.

2) The goddess of writing.

3) Thoth of Hermopolis Magna.

4) Variant *O. P.* "his hall".

5) *O. P.* obviously better "teacher of subordinates".

6) *O. P.* continues differently thus: "[skilled in (?)] his profession; knowing the secrets of heaven and earth .....; there is none who repels (?) .... in writing. First of his fellows in the midst of his neighbours; chief of his contemporaries, they are not equal [to him]; teacher of subordinates in the office of writing; his merit is proved in every stripling. Lamp ......" (end).

7) Literally: "one makes sure of him for every stripling", i. e. he is a successful teacher of the young; *iri mn m* cf. WEILL, *Sinai* 63,5. In *O. P.* this sentence more appropriately follows *sb3y n ḥṛiw-'*.

8) That *pr ḥr d-tf* is not to be taken with *hwnw nb* ("every stripling who passes through his hands") seems probable from *O. P. Ḥr* should perhaps be omitted.

9) *ip ḏ-t* "to apprehend, take stock of (one's own) body", an Egyptian phrase for "to have one's wits about one", cf. *Pyr.* 2084; *Leiden* V 93; *Urk.* IV 160. — The preposition *ḥr* is superfluous.

10) Emend *ḥr-s* referring to *s3r-t*?

11) The reading *itn* is rather uncertain.

12) Reading *spḥr*, cf. 12, 1.

13) *'Im-t*, *3bw-t* "charm of appearance" "loveliness of appearance"; cf. esp. *Anast. III* 4,12; *Ostr. Brit. Mus.* 29549. — *Twt*, a rare adjective meaning "pleasant" "delightful" or the like, cf. *Cairo, Hymn to Amon* 11,2; *twt ḥ3-t*, epithet of Ptah, MAR., *Dend.* I 39c. 71; II 57d: *twt ḳd* epithet of Osiris, ROCHEM., *Edfou* I 15, 37; I 317.

14) *Mi ir-sn*, cf. *Ostr. Brit. Mus.* 29549.

15) Variant *Ostr. Bibl. Nat.* (*O. B. N.*) "all that goes out over his limb (read "lip") is mixed with honey."

16) *Mri*, cf. BURCHARDT, *Altkanaanäische Fremdworte* (henceforth cited as BURCH.) no. 466.

17) *Sḫpr*, cf. *p iḫ n sḫpr*, PIEHL, *Inscr.* III 45. *Nfrw*, wrongly determined in *Anast. I*; cf. *Piankhi* 64; *O. B. N.* variant *ḥtri*.

18) *Mniti* cf. *Sall. II* 4,9; *Pap. Turin* 147, col. 2, 12, and possibly *Pap. Kahun* 14, 5; here metaphorically.

19) Variant O. B. N. "excellent cultivator of his position."

20) An obscure phrase thus to be rendered literally; sense perhaps "outstripping all his olde colleagues".

21) *O. B. N.*, supported in part by the Brussels ostracon, continues as follows: — "*wēb*-priest of Sekhme, excellent (?) beyond (?) men of the future (?); directing the two limits of time according to their needs (?); over the secrets of the horizons; keen in converse, never yielding (?); servant of *Wsrti*, admitted to sacred ground; not revealing (?) what he knows to the *Ḥ3w-nb*; *wēb*-priest of Sekhme, Hori son of Onnofre."

rate(?)[1] to be listened to. Thou hast parted from(?) thy papyrus to no purpose[2](??). Thou didst enter[3] knowing beforehand(?): "it is not good(?); do not(?)[4] cause it to be brought", saying: "the .... are continually ....[5] at my finger-ends(?), like a book of incantations(?) at the neck of a sick man[6]; it is continually ......, it does not tire of being fastened by the thread[7] of my ring".

VI. The author reproves Amenemope's discourtesy.

7,5 I reply to thee in like fashion[8], in a letter newly (-composed) from the first word(?)[9], down to the finis(?)[10], filled with expressions of my lips invented by myself alone, none other being with me. By the might of Thoth[11]! I made it by myself, without summoning any scribe that he might help[12]. I will give thee more than(?)[13] twenty themes(?), I will repeat for thee what thou hast said, (every) theme in its place, (from) the fourteen columns(?)[14] ⟨of⟩ thy letter. Take for thyself(?)[15] a sheet of papyrus; I will tell thee many things, and will pour out for thee choice 8,1 words[16]. — The Nile spreads abroad (its) flood when(?) the inundation season is bright(??); it takes possession of the meadows[17]. — All my words are sweet and pleasant in the saying(?)[18]; I do not act as thou dost when [I] invoke(?) thee. Thou startest with insults to me[19]; thou dost not salute me at the beginning of thy letter. Thy words are far from me, they do not come near (me); for Thoth,

---

1) *S-ẖrỉ-ʿ* seems to mean "to underestimate" (causative from *ẖrỉ-ʿ* "subordinate") in 8, 7; 17, 6. Here possibly "to think inferior". Obscure in *Pap. Turin* 146, 11.

2) Very uncertain. *G3b-tỉ* (so *O. T.*) may mean "thou art bereft of", "short of", see p. 10*, n. 17. For *ḳb* cf. p. 9*, n. 7. — *O. T.* here adds an obscure sentence.

3) The final sentences of the section are exceedingly obscure, and my translation is merely tentative. What may be meant is that Amenemope, though aware of the inferior quality of his writing, yet boasts that he has his learning at his finger-tips, and that it clings to him as closely as the magical charm round the neck of a sick man.

4) Grammatically it would be also possible to translate "it is not good that thou shouldst not cause it to be brought", *tm-k* then being infinitive (see Sethe, *Verbum* II § 587. But this gives no sense. *Tm-k* may be the *sdm-f* form optatively used, a use perhaps confined to *Anast. I*; cf. 9, 7(?); 13, 4; 28, 7.

5) *B-g* and *w-t-m* are unknown words.

6) Charms were often hung around the necks of sufferers; compare for example the directions given in Erman, *Zauberspr. f. Mutter u. Kind* 1, 4; 8, 2; 9, 3. 7.

7) *Ḥs3*, in Coptic ϩⲱⲥ, cf. *Harris I* 13b, 3; 21a, 8; 52b, 2. Spiegelberg gives an archaeological illustration *Ä. Z.* 44 (1907), 123.

8) Lit. "with the like of it", *scil.* of your letter.

9) A corrupt word.

10) *Ḳ-r*, only here.

11) Thoth is invoked as the patron deity of scribes.

12) The translation of the word *mtr* in this text often gives difficulty. The root meaning appears to be "to be present"; secondary meanings are "to bear witness", "to assist" (both in the English sense and in that of the French "assister") and even "to instruct".

13) *Dỉ-t ḥ3w n* elsewhere means "to surpass", but there are no instances of this sense with the name of a person after the preposition. The translation here given is doubtful; "more than" ought to be *ḥ3w ḥr*, not *ḥ3w m*.

14) *Sd-t*, see above p. 10*, n. 5; the construction of these words is not clear.

15) One expects *ỉ-mḥ-tw m* or *ỉ-mḥ nk m*; see *Rec. de Trav.* 27, 205.

16) *Stp-w*, cf. 2, 6; 5, 1.

17) The simile here abruptly inserted clearly alludes to the copiousness of the writer's eloquence. — The construction of *bḳ 3ḫ-t* is quite obscure. — *'I3d-t* cf. *Eloquent Peasant* B 1, 143; *Totb. ed.* Nav. 169, 5; L., *D.* III 140b, 8; *Dachel stele* 12; Mar., *Dend.* I 61a.

18) *M ḏd* usually means "to wit", and introduces a quotation; this seems hardly to be possible here.

19) Lit. "thou dost begin against me with insults"; for *sḥwrỉ* cf. p. 9*, n. 3.

my god, is a shield about me. By the might of Ptah, the Lord of Truth, ... Behold make thy words so that(?) they come to pass[1]; make 8,5 every utterance of thy mouth into every (kind of) enemy. (Yet[?]) shall I be buried in Abydos [in] the abode[2] of my father; for I am the son of Truth in the city of the Lord(?)[3] ⟨of Truth⟩. I shall be buried among my compeers(?) in the hill of the Sacred Land. Wherefore am I (as) a reprobate in thy heart? Then visit(?) it (upon me[?])[4]. To whom have I recalled thee with an evil recalling?[5] I will make for thee a volume[6] as a pleasant jest (so that) thou becomest the sport of everybody[7].

VII. Amenemope vilifies the author and is answered.

Again thou sayest (concerning me[8]): "Feeble[9] of arm, strengthless one"! Thou dost undervalue[10] me as scribe, and sayest: "He knows nothing!" — I have 9,1 not[11] spent a moment beside thee coaxing thee and saying: "Be my protector[12], someone is persecuting me[13]!" (By) the decree of the Victorious Lord, whose name is powerful, whose statutes are firmly-established like (those of) Thoth; I am the helper[14] of all my fellow-men! Thou sayest, "Thou fallest(?)", again(?) concerning me . . . . I know[15] many men without strength, feeble of arm, weak of fore-arm, lacking in power. And (yet) they are rich in houses, in food and sustenance, and they speak no wish [concerning anything(?)][16]. Come, let me tell to thee the

---

1) Note that *wꜣ r ḫpr* means "come to pass", not "are far from happening", see *Admonitions* p. 53. — If I grasp the drift of these sentences it is: "Do thy very worst, make thy words as hostile as thou wilt; yet thou canst not prevent my being buried in Abydos in the tomb of my father". It is significant that the name of the writer is Hori, and that of his father Onnofre. — The Louvre and Cailliaud ostraca (*O. L.* and *O. C.*) omit from "By the might of Ptah" down to "come to pass".

2) Variant *O. L.* and *O. C.* "tomb".

3) *Anast. I* has *n-t nb*, which can hardly here mean "every town"; I therefore take *nb* as "lord" and assume that *Mꜣʿ-t* has been erroneously omitted. *O. L.* and *O. C.* have "I am the son of Right (Truth) in the island of the Righteous" i. e. Abydos, see p. 7*, n. 1.

4) Lit. "then thou bringest it". The ostraca have *wṯsy-k wi*, which suggests a similar meaning; *wṯs* sometimes means "to inform against", "accuse", see Erman's note on *Westcar* 12, 23.

5) The text of the papyrus is corrupt; *O. L.* supplies the true reading and, together with *O. C.*, gives the variant "with evil words" for "with a bad recalling".

6) *Sḫry* seems to be the technical word for "a papyrus"; so *gstỉ ḥnʿ sḫr-t*, "palette and papyrus" *Decree of Haremheb*, 13; *sḫr-t n šʿ-t* "a papyrus-letter" *Pap. Leiden 369*; *Pap. Bibl. Nat. 198*, 1, 19; the same word too above 4, 8.

7) *O. L.* and *O. C.* substitute "people rejoice at reading it (*sdm* "to listen to" someone reading aloud, practically equivalent to our "to read") as it were a sport."

8) "Concerning me", so the ostraca.

9) Lit. "broken".

10) *Sḫrỉ-ʿ*, see above p. 11*, n. 1.

11) The ostraca have *n(ꜣ) ỉr-y* i. e. probably "have I spent?" — *n(ꜣ)* being the equivalent of *ỉn*.

12) *Mʿwnf*, see above p. 9*, n. 13.

13) Lit.: "another is hurting me". — *ꜣIꜣd* "to persecute", "hurt", actively, cf. *Totb. ed.* Leps., 162, 3; R., *I. H.* 141, 21.

14) *Psdy*, see above p. 9*, n. 16. The author seems to state that he was a helper of other men by royal appointment. The wit of this passage is not very intelligible to the Western mind.

15) Here begins a long descriptive passage where various persons with moral or physical defects are named, who yet have attained to wealth and dignities. The implication seems to be that Amenemope, the royal scribe, is only one grade better than them. Spiegelberg has dealt with the passage *Ä. Z.* 44 (1907), 123—125.

16) Cf. for example *Urk.* IV 61; here however there is not room for *r ḫt nb* in the lacuna.

middle of 30 cubits, its batter(?) 15 cubits, its base(??) of 5 cubits. The quantity of bricks needed for it is asked of the commander of the army. All the scribes 14,5 together lack knowledge among them(?). They put their faith[1] in thee, all of them, saying: "Thou art a clever scribe[2], my friend! Decide for us quickly! Behold thy name is famous; let one be found in this place (able) to magnify[3] the other thirty! Let it not be said of thee that there is aught that thou dost not know! Answer us (as to) the quantity of bricks needed! Behold its measurements(??)[4] are before thee; each one of its compartments(?) is of 30 cubits (long) and 7 cubits broad"[5].

XIII. Amenemope is unable to determine the number of men required in the transport of an obelisk of given dimensions.

15,1 Come, good sir, vigilant scribe, who art at the head of the army, distinguished when thou standest at the great Palace-gates, comely when thou bowest down beneath the Balcony[6]! A despatch has come from the crown-prince[7] at *Rʿ-kꜣ*[8] to rejoice the heart of the Horus of Gold, to extol(?) the raging Lion(?)[9], telling that an obelisk[10] has been newly made, graven with the name of His Majesty, of 110 cubits in length of shaft; its pedestal 10 cubits (square), the block at its base making 7 cubits in every direction; it goes in a slope(?) towards the 15,5 summit(?), one cubit and one finger(?); its pyramidion one cubit in height, its point(?) (measuring) two fingers. Add them together(??) so as to make them into a list(??)[11], so that thou mayest appoint every man needed to(??) drag them, and send them to the Red Mountain[12]. Behold, they are waiting for

1) *Mḥ ib* only here exactly in this sense; but the phrase is used of having confidence in something asserted or believed; see my *Inscription of Mes* p. 15, n. 23.

2) Variant *P. T.* "thou art keen of wit".

3) I. e. able, by solving the problem, to save the reputation of his colleagues. — *Gm* with object, closely followed by *r* with an infinitive, is an idiom with various slight shades of meaning e. g. "to find someone able", "ready", "competent" to do something. Cf. below 23, 2; 28, 8; *Pap. Leiden* 370, recto 16; *Anast. V* 9, 4; 17, 7; *Turin Lovesongs* 1, 14.

4) *Ḫtiw* might mean either (1) "pedestals", "steps", or (2) "threshing-floors"; neither of these senses is here suitable. What we clearly need is a word for "measurements", and perhaps the original reading was *ny-f ḫꜣy* "its measurements"; for *ḫꜣy* cf. Daressy, *Ostraca* 25262 (Cairo); *Pap. Turin* 71, 1 (omitted in facsimile).

5) Amenemope makes no answer, and the subject is dismissed in silence.

6) *Sšd*, ϣⲟⲩϣⲧ *fenestra*; see Hölscher, *Das hohe Tor von Medinet Habu*, p. 49—50 for a good archaeological illustration.

7) *Rpʿti* in the New Kingdom is as a rule the designation of the Crown-prince; convincing examples are *d'Orbiney* 19, 2. 6; *Harris I* 42, 8; *Pap. Turin* 17, 1, 102, 2, 9; *Inscr. dédic.* 44.

8) *Rʿ* probably here means "district of" or the like. A canal or branch of the Nile named *Kꜣ* is mentioned on several wine-jars from the Ramesseum, viz. Spiegelberg, *Hier. Ostr.* nos. 209. 217. 218. 269. 289. 292. In the last-quoted instance *Kꜣ* is connected with "the water of Ptah", which is found in a list of canals etc. appended to the Catalogue of Lower Egyptian nomes (see Brugsch, *Dict. Géogr.* 239); but that the word *kꜣ* in the same list (*op. cit.* 1271) is no geographical name seems fairly clear. For the location of *Kꜣ* note that the Red Mountain is implied in 15, 6 to have been the quarry where the obelisk was made; this is the Gebel Aḥmar near Cairo, see my *Notes on the Story of Sinuhe*, on B 14—15.

9) I. e. the Pharaoh. Probably we should read *r swh mꜣi nšn*; for *swh* with a direct object see *Admonitions* p. 28.

10) For the technical words see the Appendix.

11) A very difficult and uncertain sentence.

12) See above n. 8.

[them][1]. Prepare(?) the way for(?) the crown-prince *Ms-ỉtn*. Approach(?)[2] and decide for us the number of men who (shall go) before him. Let them not have to write again! The monument (lies ready) in the quarry. Answer quickly, do not dawdle[3]! Behold thou art seeking them[4] for thyself! Get thee on[5]! Behold 16,1 thou art bestirring thyself(?)[6]. I cause thee to rejoice; I used formerly to ...... like thee. Let us join the fray together[7], for my heart is tried, my fingers are apt and clever[8] when thou goest astray. Get thee (onwards)[9]! Do not weep! Thy helper[10] stands behind thee! I will cause thee to say: "There is a royal scribe with the Horus, the Victorious Bull", and thou shalt order men to make chests into which to put letters[11]. I would have written for thee stealthily(??)[12], 16,5 but(?) behold thou art seeking it for thyself[13]. Thou settest my fingers ....[14] like a bull at a festival at every festival of ....

XIV. Amenemope proves himself incapable of supervising the erection of a colossus.

It is said to thee: "Empty[15] the magazine that has been loaded with sand under the monument of thy Lord which has been brought from the Red Mountain. It makes 30 cubits stretched upon the ground, and 20 cubits in breadth, ... -ed with 100(??) chambers[16] filled with sand from the river-bank. The .... of its(?) chambers have a breadth of 44(?) cubits and a height of 50 cubits, all 17,1 of them, ... in their ...."[17] Thou art commanded to find out what is before (the Pharaoh)(??)[18]. How many[19] men will (it take to) demolish

1) *Sỉn* (written like, but a totally different word from, *sỉn* "to hasten") means "to wait" cf. *Sinuhe R* 21; *Sphinx stele* 11; *sỉn n* "to wait for" cf. *d'Orbiney* 3, 1; *Anast. IV* 5, 1; *Pap. Turin* 136, 2; 68, col. 3, 3. 12; *Pap. Leiden 345* verso G 4, 2. 3. 4. 7.

2) If the sentences are here rightly divided, *ỉmỉ w3-t* must mean "prepare the way" for the crown-prince, who would be unable to start without the men who are to drag the obelisk from the quarry. But it is not certain that *ẖn* "approach" is here an imperative; *n* might be equivalent to *ỉn*, and *ẖn* predicate "make way, the crown-prince approaches". But the meaning would then be very obscure.

3) *'Intnt* "to linger", "to hesitate", see *Ä. Z.* 45, (1908), 61.

4) Them, i. e. the number of men required; or, the solution of the problem.

5) *Ms* reflexively *Westcar* 10, 12; *Turin, statue of Haremheb* 15; R., *I. H.* 223 = *Sall. III* 7, 6; in the imperative as here, cf. *Pyr.* 586. 645. 1657; Rochem., *Edfou* II pl. 30 c; L., *D.* IV 57 a.

6) Read *ms-k tw*(?); for similar corruptions see the critical note p. 36a, note 5f.

7) The writer appears to be offering his help, alleging that he himself was once in similar straits and therefore knows how to cope with such difficulties. *Ṯs skw*, see *Admonitions* p. 20; *n sp* is the Coptic ⲛ̄ⲥⲟⲡ.

8) Lit. "hear (understand) cleverness".

9) Erman rightly emends *ỉms-tw* as in 15, 8.

10) For *'ḏr* see above p. 9*, n. 14.

11) The meaning perhaps is that Amenemope, having found a helper, not only loudly exclaims that Pharaoh possesses in himself a competent royal scribe, but even goes so far as to order the boxes into which his letters are to be put. — The suffix *-sn* makes it necessary to emend the plural *pdsw* "boxes".

12) *H-r-ṯ-ṯ* only here.

13) See above 15, 8, and n. 4 on this page.

14) *Nsns*, ἅπαξ λεγόμενον. Here again the sense is utterly obscure.

15) For the mode of erection contemplated see the Appendix.

16) *Šmm* see above p. 9*, n. 2.

17) *Ḏ3y*, *twḥw* (?) and *sg3* are unknown words.

18) It is hardly possible to translate differently, but my rendering gives no satisfactory sense. *M b3ḥ* seems to be used in reference to ascertained dimensions in 14, 8.

19) *Wr*, Coptic ⲟⲩⲏⲣ, cf. below 27, 8; *Anast. V* 20, 5; *Unamon* I, x + 15.

it[1] in six hours — (if[?])[2] apt are their minds(?), but small their desire to demolish it without there coming a pause when thou givest a rest[3] to the soldiers, that they may take their meal[4] — so that the monument may be established in its place? It is Pharaoh's desire to see it beautiful!

XV. Amenemope fails to make proper provision for a military expedition.

O scribe, keen of ⟨wit⟩, understanding of heart[5], to whom nothing whatsoever is unknown, flame[6] in the darkness before the soldiers, giving light to them! Thou art sent on an expedition to Phoenicia(?)[7] at the head of the victorious army, in order to smite those rebels who are called Neârîn[8]. The troops of soldiers who are before thee amount to 1900; (of) Sherden 520(?), of Kehek 7,5 1600, of Meshwesh ⟨100(?)⟩[9], Negroes making 880; total 5000 in all, not counting[10] their officers. A complimentary gift[11] has been brought for thee (and set) before thee, bread and cattle and wine. The number of men is too great for thee, the provision[12] (made) is too small for them: loaves of ..... flour[13], 300; cakes[14], 1800; goats of various sorts, 120; wine, 30 (measures). The soldiers are too numerous, the provisions are underrated[15] as compared with(??) that which thou takest of them. Thou receivest(?) (them, and) they are placed in the camp. The soldiers are prepared and ready. Register them quickly, the share of[16] every man to his hand. The Beduins look on in secret[17]. O sapient scribe[18], midday has come, the camp is hot. They say[19]: "It is time to start[20]! Do not make the commander[21] angry! Long is the march before us!" But I say: "What means it, that there is no(??) 8,1 bread at all[22]? Our night-quarters are far off! What means, good sir, this scour-

---

1) *Ḥm* doubtless originally "to demolish" a wall; "to force open" a tomb, cf. *Pap. Amherst* 2, 2; *Mayer B* 9: elsewhere chiefly metaphorically, e. g. *Pyr.* 311. Cf. Coptic ϭⲟⲙϭⲉⲙ *confringere*.

2) The difficult words that follow seem to refer to the likelihood that the workers, though competent, will show themselves unwilling to work for six hours continuously without a break for a meal; in calculating the number of men required this factor must be taken into account.

3) *Rdìt srf n*, *Sphinx stele* 6; Fraser, *Scarabs* 263, 14; *Decree of Haremheb* 25; *ibid.* right side 10; cf. ⲥⲣϥⲉ *otiari*.

4) The word *ʿš* only here.

5) *Wḥʿ ìb*, see *Ä. Z.* 45 (1909), 136.

6) For the metaphor cf. L., *D.* II 150a, 4 (Hammamat).

7) All translators have here emended *Rhn*, Hammamat, but the context demands the name of a well-known country in the direction of Syria. For my conjecture *Ḏ3hì* see the note on the reading p. 29a, note 1b.

8) *N-ʿ-r-n*, נערים "warriors", a sense that is found in the O. T. See Burch. no. 559.

9) The Meshwesh and the Negroes are never elsewhere linked together as the Ms. reading suggests; hence a number may have fallen out of the text. See p. 29a, note 4c.

10) *W3-tw*, see *Ä. Z.* 47 (1910), 134—136.

11) From the stem שלם; Burch. no. 871.

12) *Nkt* "provisions", an exceptional sense of the word found again *Anast. IV* 13, 12.

13) *Ḳmḥ*, an ancient Egyptian word related to Hebrew קמח; Burch. no. 984.

14) *'I-p-t*, from אפה "to bake", Burch. no. 39.

15) *S-ḫrì-ʿ*, see above p. 11*, n. 1.

16) *P-n*, see above p. 13*, n. 8.

17) *M ṯ3w(t)*, Coptic ⲛ̄ϫⲓⲟⲩⲉ *furtim*, cf. below 20, 4; *Anast. IV* 4, 11.

18) Two Semitic words, which in Hebrew would be ספר יודע.

19) The reading is uncertain; see p. 29a, note 16i.

20) *F3* "to start", cf. 20, 1.

21) *Ṯs pd-t*, cf. Mar., *Abydos* I 53; Ros., *Mon. Stor.* I 125.

22) *M kf* is elsewhere found at the end of negative sentences for emphasis (like *ìn* ⲁⲛ), cf. 27, 3; *Anast. IV* 13, 5; *Anast. V* 7, 1; 17, 7. Hence the conjecture *bn* here.

a halt in the evening[1]; all thy body is crushed and battered(?)[2]; thy [limbs]
20,1 are bruised(?)[3] . . . . from sleep[4]. Thou wakest, and it is the hour for starting[5] in the drear(?) night. Thou art alone to harness (the horse); brother comes not to brother. A fugitive(??)[6] has entered into the camp. The horse has been let loose[7]. The . . . . has turned back(?)[8] in the night. Thy clothes have been taken away. Thy groom has awoke in the night, and marked what he has done(?)[9]; he takes what remains and joins (the ranks of) the wicked, he mingles with the people of the Shosu and disguises himself[10] as an Asiatic.
20,5 The enemy comes to pillage[11] in secret. They find thee inert. Thou wakest up and findest no trace of them[12]; they have made away[13] with thy things. Thou art becoming a fully-equipped[14] Maher, thou fillest thy ear(?)[15].

XVII. The Phoenician cities.

I will tell thee of another mysterious city. Byblos[16] is its name; what is it like — and its(?) goddess, once again? Thou hast not trodden it. Come teach me[17] about Berytus[18], and about Sidon[19] and Sarepta[20]. Where is the

---

1) Lit. "proceedest to stop"; for *spr* as an auxiliary verb cf. *Anast. IV* 9, 11; *Anast. V* 10, 7. — *Wḥ*ʿ perhaps properly "to stop", "leave off" work, so *Paheri* 3; derivatively, "to return" from work, so *d'Orbiney* 4, 3; R., *I. H.* 248, 85: *wḥ*ʿ *m rwhȝ* also, in a somewhat similar sense, *d'Orbiney* 4, 7; 13, 7.

2) *Hdhd*, only here in this sense; cf. *Urk.* IV 710 for *ḥdhd* in a quite different sense.

3) *Wš(ȝ)wš(ȝ)*, cf. Boh. ⲟⲩⲉϣⲟⲩⲱϣ-, properly "to bruise" "crush", cf. *Anast. IV* 9, 7 = *Anast. III* 5, 9; *Anast. V* 10, 7 = *Sall. I* 3, 9; derivatively "to break" "smash", see below 26, 1; then "to break open", *Mayer A*, recto 3, 4; unpublished *Turin papyrus* = Spieg., *Zwei Beiträge*, p. 12.

4) I now believe that the word *ṯnm* (sic?) is on a misplaced fragment. The restoration of the passage is quite obscure.

5) *Fȝ* in this sense, see above 17, 8.

6) *N-h-r* elsewhere (R., *I. H.* 143, 41) means "to flee" and is probably connected with Semitic נהר "to flow". Here the feminine article *tȝ* is incomprehensible, and the sense is obscure.

7) *Ṯt* "to untie" "loose", cf. *Pap. Turin* 23, 6; 33, 9; 73, 10; *Israel stele* 6; *Harris 500* recto 4, 8; *Vatican Magical Pap.* = *Ä. Z.* 31 (1893), 122.

8) *Ḫtḫt-(tw)*, the pseudoparticiple **ḫetḫōt*; it is tempting to render "has been ransacked", comparing ϩⲟⲧϩⲧ : ϭⲟⲧϭⲉⲧ, but no evidence for this meaning is forthcoming in late-Egyptian.

9) *'Ir-nf* can hardly be meant for "what has been done to him", cf. 18, 7; the sense is obscure.

10) See above p. 14*, n. 15.

11) *Š(ȝ)d(ȝ)*, probably שדד "to be violent", "to violate", see Burch. no. 893.

12) For ʿ "trace" cf. *Anast. V* 20, 4; obscure L., *D. III* 140d, 5.

13) *Rmn* only once again in this sense, *Urk.* III 106; cf. *mnmn*.

14) *Sdbḥ* "to furnish" "equip", cf. *Anast. IV* 12, 6 = *Anast. V* 3; *Anast. IV* 13, 10 = *Koller* 5, 8; *Anast. IV* 16 verso, 6; *Harris I* 77, 9.

15) The *m* of *mḥ-k m msḏr-k* is perhaps to be omitted; see *Koller* 3, 2. The sense may perhaps be: thou art listening attentively, and gradually acquiring the experience of a Maher.

16) For Byblos and its goddess Hathor see especially Sethe's article, *Ä. Z.* 45 (1908), 7—14. — The list of Phoenician towns is in correct geographical order from North to South.

17) *Mtr-i my r* also 21, 5; 22, 1. 7; the infinitive here probably replaces the imperative, see Sethe, *Verbum* II § 566. *Mtr r* "to instruct" someone about something, cf. *Anast. IV* 14, 8; *Decree of Haremheb*, right side, 5.

18) *B-r-t*, Berytus, *be-ru-ta* in the Amarna tablets; only here in Egyptian texts, see Burch. no. 366.

19) *Ḏ-d-n*, Sidon, צידון, mentioned also *Unamon* 1, x + 24.

20) *Ḏ-r-p-t*, צרפת, Σαρεπτα, Sariptu of the Taylor cylinder of Sennacherib, on the high road between Tyre and Sidon. In Egyptian only here.

21,1 stream of *N-ṯ-n*[1]? What is *'I-ṯ*[2] like? They tell of another city in the sea, Tyre-the-port[3] is its name. Water is taken over to it in boats, and it is richer in fishes than in sand.

XVIII. Places further southwards.

I will tell to thee another misery[4] — the crossing of *D̠-r-ʿ-m*[5]. Thou wilt say: "It burns more than a (hornet-)sting[6]!" How ill it goes with the Maher! Come, set me on the road southward to the region of Acco(?)[7]. Where is the 21,5 road of Achshaph[8]? Beside(?) what city (does it pass)? Pray teach me about the mountain of *Wsr*[9]; what is its peak like? Where is the mountain of Shechem[10]? Who ....? The Maher — where does he make the journey to Hazor[11]? What is its stream like? Put me ⟨on⟩ the route to *Ḥ-m-t*[12], *D-g-r*[9] 22,1 and *D-g-r-êl*, the playground[13] of all Mahers. Pray teach me about his road. Make me behold *Y-ʿ-n-...*[9]! If one is travelling to *'I-d-m-m*[14], whither turns the face? Do not make ⟨me(?)⟩ withdraw(?)[15] from thy teaching, lead me(?) to know them!

---

1) The stream of *N-ṯ-n* can only be the Nahr el Kasîmîye, i. e. the lower courses of the Lîtânî, see MASPERO, *Hist. Anc.*, II p. 6, note 6; the identification of the name *N-ṯ-n* with Lîtânî is however open to serious objections, see MAX MÜLLER, *As. u. Eur.* 185.

2) *'I-ṯ* (BURCH. no. 190) is Uzu in the Tyrian series of the Amarna letters; ED. MEYER, *Encycl. Bibl.* col. 3733, accepts PRAŠEK's identification with Palaetyrus.

3) *D̠-r*, often in Egyptian (BURCH. no. 1227), is צֹר, Τύρος. The island of Tyre lies about a mile from the shore and lacks both water and vegetation. For the translation "Tyre-the-Port", see MAX MÜLLER, *As. u. Eur.* 185, note 1.

4) *Tp-ḳsn* only here; but *tp* is used with various adjectives in a similar way, cf. *tp-nfr*, *tp-mtr*, *tp-wʿ*, *tp-šw*.

5) The locality *D̠-r-ʿ-m* (the final *m* may well be a corruption of *n*[ꜣ]) is compared by MAX MÜLLER with צרעה Σαραα of *Judges* 18, 2; *Joshua* 19, 41, which was in Dan. This seems too far south for the context. DE ROUGÉ (quoted by BRUGSCH in the *Critique*) cleverly suggested that there is here a pun upon the word צִרְעָה "hornets"; see next note.

6) *D̠db* "to sting" is not elsewhere determined with the sign for fire; but *d̠dm*, demonstrably only another form of the same word, is so determined *Pap. Turin* 133, 12. Thus DE ROUGÉ's ingenious suggestion mentioned in the last note is brilliantly confirmed.

7) The word "southward" shows that the writer is following, or at least intends to follow, some geographical order. Almost immediately after Tyre no place could more appropriately named than Acco, for which *ʿ-k-n* is doubtless a corruption. All the known localities in this section lie to the S. of Carmel with the exception of Hazor.

8) It is probable that *ʿ-k-s-p* is a misspelling of *i-k-s-p* (*Urk.* IV 782, 40), i. e. אַכְשָׁף on the border of Asher; see BURCH. no. 168. E. MEYER (*Encycl. Bibl.* 3733) identifies *ʿ-k-s-p* with אכזיב; this is a particularly attractive suggestion, since Achzib is in the near neighbourhood of Acco; the equivalence *s* = ז is however open to serious objections.

9) Unknown name.

10) *S-k-m*, evidently the שְׁכֶם of the old Testament, see MAX MÜLLER, *As. u. Eur.* 394; probably it is Mount Ebal that is meant.

11) *Ḥ-d̠-r* clearly corresponds to Hebrew חצור (BURCH. no. 709) and to *Ḫa-zu-ri* of the Amarna Tablets. This town was situated near the waters of Merom, not far from Kedesh.

12) *Ḥ-m-t* here and in the Palestine list of Thutmosis III is thought to be Hammath חמת south of the Sea of Galilee, see MAX MÜLLER, *Die Palästinaliste Thutm. III*, p. 11; BURCH. no. 678.

13) *T is-t swtwt* lit. "the place of promenading"; *swtwt* means "to walk for pleasure" "to promenade" "make an excursion".

14) *'I-d-m-m* (cf. *'I-t-m-m* in the list of Th. III) has been compared with the בְּצַלַּח אֲדֻמִּים on the border between Benjamin and Judah; see MAX MÜLLER, *Die Palästinaliste Thutm. III*, p. 15.

15) Utterly corrupt; emend *sḥnḥn-t* (?).

XIX. Various other towns visited by the Maher.

Come let me tell thee of other towns, which are above(??)[1] them. Thou hast not gone to the land of *T-ḫ-s*[2], *K-w-r-m-r-n*[3], *T-m-n-t*[4], Kadesh[5], *D-p-r*[6], 22,5 *I-ḏ-y*[8], *H-r-n-m*[8]. Thou hast not beheld Kirjath-anab and Beth-Sepher[7]. Thou dost not know *'I-d-r-n*[8], nor yet *Ḏ-d-p-t*[9]. Thou dost not know the name of *Ḫ-n(r)-ḏ*[10] which is in the land of Upe[11], a bull upon its boundary, the scene of the battles of every warrior. Pray teach me concerning the appearance(?) of *Ḳ-y-n*[12]; acquaint me with Rehob[13]; explain Beth-sha-ēl[14] and *T-r-ḳ-ēl*[15]. The stream 23,1 of Jordan[16], how is it crossed?

Cause me to know[17] the way of crossing over to Megiddo which is above it(??)[18] Thou art a Maher skilled in the deeds of the brave[19]! A Maher such as thou art is found (able) to march(?)[20] at the head of an army! O

1) *Ḥry* is perhaps corrupted from the form of the preposition *ḥr* used before the suffixes *-tn* and *-sn*; a difficult phrase *nti ḥr-f* again below 23, 1. — The places here mentioned appear to range from the North of Syria to the extreme South of Palestine.

2) *T-ḫ-s* is very frequent in Egyptian texts; *Taḫ-si* of the Amarna letters, where it is mentioned together with the land of Ube, see BURCH. no. 1128.

3) Unknown name.

4) *T-m-n-t* is surely not תמנת in Judah; for while of the seven names here given four are unknown, the other three are N. of Damascus.

5) *Ḳdš* is Kadesh on the Orontes, see especially BREASTED, *The Battle of Kadesh*, 13—21. A Kadesh was mentioned above in 19, 1, see p. 21*, n. 5.

6) *D-p-r*, a town stormed by Rameses II, probably quite close to Kadesh, see MAX MÜLLER, *As. u. Eur.* 221; BREASTED (*Ancient Records* III 159) places it further south.

7) The Ms. has Kirjath-*ʿ-n-b* and Beth-*ṯ-p-r*. MAX MÜLLER (*As. u. Eur.* 170) formerly proposed to interchange Kirjath and Beth in these names, since קִרְיַת-סֵפֶר (so LXX rightly) is mentioned in conjunction with עֲנָב in *Joshua* 15, 49. However DARESSY reads Kirjath-*ʿ-n-b(w)* in a list of foreign names at Abydos (*Rec. de Trav.* 21, 2), where MARIETTE read Kirjath-*ʿ-n-t(w)*; and MAX MÜLLER himself has found Kirjath-*ʿ-n-b-w* in a palimpsest list at Karnak (*Researches* I 57, 14). There can be little doubt, in any case, that the same places are meant as are referred to the passage of Joshua; these are situated in the hill-country of Judah.

8) For *i-d-r-n* the *i-d-r-m* of the list of Sheshonk (19) and אֲדוֹרַיִם have been compared, see BURCH. no. 201. The Adoraim of the O. T. is identified with modern Dūra, to the S. W. of Hebron in Southern Judaea.

9) Cf. the name *Ḏ-d-p-t-r* in the Sheshonk list (34); otherwise unknown.

10) Unidentified.

11) See above p. 20*, n. 12.

12) *Ḳ-y-n* is very probably identical with *Ḳn* in the Annals of Thutmosis III (*Urk.* IV 655. 657), near Megiddo, the *Gina* of the Amarna letters; see MAX MÜLLER in *Encycl. Bibl.* col. 3547. — The injured word for "appearance" (or "statue"??) is read *sḏi* by BRUGSCH and compared, probably wrongly, with the late word *sḏḏ* (*Wörterb.* 1357).

13) *R-ḥ-b* is רְחוֹב in Asher, often mentioned in Egyptian texts; see BURCH. no. 628, and MAX MÜLLER, *As. u. Eur.* 153.

14) *Byt-š-ir*, a בת-ש-אל, often named in hieroglyphs, see BURCH. no. 388; not localised, but it occurs next Rehob, as here, in the Sheshonk list (16).

15) Unknown; it is tempting to transpose and read *ḳ-r-t-ir* i. e. a קרית-אל; so CHABAS, but see MAX MÜLLER, *As. u. Eur.* 175, footnote.

16) *Y-r-d-n*, in Hebrew ירדן, only here; the word ought to have the determinative of water.

17) See the critical note.

18) *Nti ḥr-f* can hardly be translated "which is upon it", since Megiddo is not on the Jordan, while the writer's knowledge of Palestinian geography was evidently admirable. On the other hand to render "qui est en outre de cela" (CHABAS), comparing *nti ḥry-sn* 22, 3, is a very bold expedient.

19) *Pr-ʿ-ib*, cf. 26, 9; *Anast. II* 3, 6; *Sall. II* 10, 1. 9.

20) *S-g* here only. — For the idiom *gm r* see above p. 17*, n. 3.

Mariannu[1], forward to shoot(?)[2]! Behold the . . . . . .[3] is in a ravine[4] two thousand cubits deep, filled with boulders[5] and pebbles[6]. Thou drawest back(?)[7], thou graspest the bow, thou dost . . . .[8] thy left hand, thou causest the great 23,5 ones to look. Their eyes are good, thy hand grows weak(?)[9]. אבדת כמ ארי מהִר נעם[10]. Thou makest the name of every Maher, officers of the land of Egypt[11]. Thy name becomes like (that of) *K-ḏ-r-d-y*, the chief of *'I-s-r*[12], when the hyena[13] found him in the balsam-tree[14]. — The(?) narrow defile[15] is infested(?) with Shosu concealed beneath the bushes; some of them are of four cubits or of five cubits, from head(??) to foot(?)[16], fierce of face, their heart is not mild, and they hearken not to coaxing. Thou art alone, there is no helper(?)[17] with thee, no army[18] 24,1 behind thee. Thou findest no . . . . . .[19] to make for thee a way of crossing.

---

1) *M-r-y-n*, a word found often in Egyptian texts in reference to Syrian "warriors" (Burch. no. 470). The translation "lords" seems to me erroneous, though doubtless the warrior-class was held in high honour in the small Syrian states. This translation is probably due to the old etymology from Aramaic מריא (Chabas); another Semitic derivation that has been proposed is from מרה "to be contentious" "rebellious". Winckler has recently found the word in the form *mariannu* in the tablets from Boghazkoï, and boldly connects the word with the Vedic *márya* "man" "hero" (*Or. Lit. Zeit.*, 13 [1910], 291—298). Both here and in 28, 1 it is apparently parallel to *m-h-r* (Maher).

2) *N ḥr-k*, cf. 15, 8; 24, 1; cf. too Düm., *Hist. Inschr.* II 47, 4 and the passages quoted by Griffith, *Proc. S. B. A.* 19, 298. *N ḥr-tn rs* is used as an exclamation "forward!" in *Piankhi* 95.

3) A corrupt word with the determinative of land; probably from the stem *n*ʿʿ. — It seems necessary to delete the preposition *ḥr* after *mtk*.

4) *Š-d-r-t* only here and in 24, 3; the approximate sense seems certain.

5) *Dḥ-wt* only here and 24, 2.

6) The word ʿ*-n(r)* is the Coptic ⲁⲗ, see Burch. no. 270. 274.

7) *S-w-b-b* is thought to be שׁובב "to turn back" (Burch. no. 768), though neither determinative nor sense seems very appropriate. The word cannot be identified with סבב, as Chabas supposed.

8) *P-r-ṯ*, an unknown verb.

9) *Wrw nfr* (sic) is usually divided from what follows; thus Erman translated "so ermüdet ihr Auge auf deiner Hand". But *gnn* is not used of the eye, nor does it mean "to grow tired"; and the preposition *ḥr* would be strangely used. Probably *ḥr* should be omitted; for *gnn dt-k* cf. *Koller* 5, 3; R., *I. H.* 241, 43.

10) So Burchardt (under no. 32), who proposes as the meaning of these words "thou slayest like a lion, o . . . . Maher". Whether the remainder of this translation be correct or not, it seems probable, in spite of Burchardt's objections, that the last word is נָעִם "pleasant" "delightful".

11) Without emending it is impossible to translate otherwise; *nb* cannot here mean "lord".

12) *'I-s-r* (mentioned among South-Palestinian names under Sethos I, L., *D.* III 140a) is probably the tribal name אָשֵׁר; see E. Meyer, *Die Israeliten und ihre Nachbarstämme*, p. 540.

13) *Ḥtm-t*, see above p. 21*, n. 12.

14) *Bkì*, Hebrew בכא, only here (Burch. no. 374).

15) Hitherto it has been customary to connect *ḥr tꜣ gꜣw-t* and what follows with the preceding simile, which then only ends with *swnwn-w* in 23, 8. This view is untenable for several reasons: (1) the first sentence of the simile ("when the hyena found him in the balsam-tree in the narrow defile, infested with Shosu concealed beneath the branches") is breathlessly long and gives no good sense; the presence of the Shosu would diminish, rather than increase, the dangerousness of the hyena; (2) the words "some of them" are incompatible with the singular word "the hyena" in the text of *Anast. I.* The difficulty is at once solved by the omission of *ḥr* before *tꜣ gꜣw-t*; from this point onwards it is the journey of the Mahar in the mountainous pass (*gꜣw-t* also in the sequel 24, 6) that is described, not the adventures of the prince of Asher. The statement "some were of four cubits or five cubits" (i. e. 6 foot 10 inches to 8 foot 6 inches) now refers to the Shosu, as with due allowance for Egyptian exaggeration it well may do; and the words "they do not listen to coaxing" obtain a more natural and less metaphorical meaning.

16) Whether the suggestion *fnd* in the critical note is correct remains quite uncertain. For the (collective?) form *rd-yt* I have no parallel.

17) For *ḏ-r* the context clearly demands some word for "helper", and I propose ʿ*-ḏ-r*, comparing 16, 2; צִיר "messenger" is not appropriate.

18) *Ḏ-b-ì*, the Hebrew צָבָא; cf. 27, 1 and Burch. no. 1207.

19) *'I-r-ì-r*, an unknown word; the sense required is "guide" or the like. Burch. no. 92 differently.

Thy . . . . . begs the . . . . thy mouth[1]: "Give (me) food and water, for I have arrived safely". They turn a deaf ear, they do not listen, they do not heed[2] thy tales. Thou makest thy way into the armoury[3]; workshops surround thee[4]; smiths 26,5 and leather-workers[5] are all about thee. They do all that thou wishest. They attend to thy chariot, so that it may cease from lying idle. Thy pole[6] is newly shaped(?)[7], its . . . .[8] are adjusted. They give leather coverings(?)[9] to thy collar-piece(?)[10] . . . They supply[11] thy yoke. They adjust(?) thy . . . .[12] (worked) with the chisel(?)[13] to(?) the . . . .[14] They give a . . . . (of metal)[15] to thy whip[16]; they fasten [to] it lashes[17]. Forth thou goest quickly to fight on the open field, to accomplish the deeds of the brave[18]!

XX. The first stations on the Syrian high-road. End of the Controversy. Conclusion.

27,1 Good sir, thou honoured scribe, Maher cunning of hand, at the head of the troops[19], in front of the army[20], [I will describe to] thee the [lands] of the extremity of the land of Canaan[21]. Thou answerest me neither good nor evil; thou returnest me no report. Come I will tell thee [of many things(??)]; ⟨turn(?)⟩

---

1) Difficult and corrupt words.

2) *Hn* "to heed", see Sethe, *Die Einsetzung des Veziers*, p. 21, note 91.

3) *Ḫpš*, see above p. 13*, n. 7.

4) *Ḳd* "to surround", cf. especially *Anast. IV* 12, 4.

5) *Ṯb-w* properly "sandal-makers".

6) The ʿ is certainly the "pole" of the chariot (also in the case of a single-horse chariot doubtless the double shafts); for the pole particularly good wood was selected, cf. *Koller* 2, 1; the pole comes from Upe, *Anast. IV* 16, 11; a chariot is bought, "its pole (ʿ) for 3 *dbn*, the chariot (itself) for 5 *dbn*", *Anast. III* 6, 7.

7) *G-r-p* occurs only here, but is certainly identical with *g-r-b*, *Anast. IV* 16, 11 = *Koller* 2, 1, also in reference to the chariot-pole. In Aramaic גלב is a knife for cutting, in Phoenician a barber.

8) *Dby-wt*, only here; sense unknown.

9) *M-š-y*, again only in *Koller* 2, 1.

10) *D-t*, see above p. 26*, n. 14. The following word *ḥ3w* is quite obscure.

11) For *ʿrʿr* cf. *Pap. Turin* 67, 10; *Unamon* 2, 42; R., *I. H.* 201, 8; metaphorically "to acomplish" plans, commands, cf. *Unamon* 2, 32; R., *I. H.* 145, 59; *Pap. Bibl. Nat.* 197, 3, 4. 6.

12) The *ḏbw* must be an important part of the chariot; it might be of gold, see *Urk.* IV 663. 669.

13) The usual phrase for "worked with the chisel" is *ṯ3(w) m bsn-t*, cf. *Harris I* 6, 7. 9; 47, 3. 4: *Ṯ3(w) m ṯ3 bsn* occurs *Koller* 1, 7, where as in the present passage it is hard to explain.

14) *M-ḫ-t* occurs (with determinative of wood) once again *Anast. IV* 16, 12, where it is a part of the chariot adorned with metal; the meaning is unknown.

15) *'I-ṯ-m-y*, only here.

16) *'I-s-b-r*, see Burch. no. 134.

17) *M-t-ḏ-t*, only here.

18) *Pr-ʿ-ib*, see above p. 24*, n. 19.

19) *N-ʿ-r-n*, see above p. 19*, n. 8.

20) *Ḏ-b-t*, see above p. 25*, n. 18.

21) *T3 n p K-nʿ-n*, in Hebrew ארץ כנען, only here with *t3 n*. *P K-nʿ-n* is not very often mentioned in the Egyptian texts, see Burch. no. 988; except in *Anast. III* 8, 5 = *Anast. IV* 16, 4 it has always the definite article. So far as the Egyptian texts are concerned, Canaan might be the name of merely the south of Philistia; but *Kinaḫḫi* in the Amarna letters appears to indicate a wider extension. The present passage describes the localities lying between the fortress-town of Zaru and the Philistine city of Gaza, and is strikingly illustrated by the scenes on the N. Wall of Karnak depicting the conquests of Sethos I (L., *D. III* 128b; 128a; 127a; 126b in this order). The accuracy of the author's geographical knowledge is convincingly attested by a comparison with these sculptures.

thy face(?) ⟨towards(?)⟩[1] the fortress of the "Ways of Horus"[2]. I begin for thee with the "House of Sese"[3]. Thou hast never trodden it; thou hast not eaten the fish of (the waters of) . . . .;, thou hast not bathed in them. Come prithee[4] 27,5 let me recount to thee *Ḥ-ṯ-y-n*[5]; where is its fortress? Come let me tell thee about the district of Buto of Sese[6], "In(?) his house of victories(?) of Usimarê"[7], *S-b-ēl*[8] and *'Ib-s-ḳb*[9]. Let me describe to thee the manner of *'-y-n-n*[10]; thou knowest not its position[11]. *N-ḫ-s*[12] and *Ḥ-b-r-t*[13], thou hast never seen them since thy birth. O Mohar, where is Raphia[14]? What is its wall like? How many leagues[15] march is it to Gaza[16]? Answer quickly! Render me a report, that I may call thee a 28,1 Maher, that I may boast to others of thy name of Mariannu[17]. So will I say to them(?). Thou art angry at the thing I [have] said to thee. I am experienced in every rank[18]. My father taught me, he knew and instructed(??) (me) very often. I know how to hold the reins[19], beyond thy skill indeed! There is no

---

1) The text here is damaged and probably also corrupt. The *crux* of the passage is to determine the grammatical construction of the words *p ḫtm n W3-wt Ḥr*.

2) *W3-wt* [*Ḥr*] is now known to be an alternative name for the celebrated frontier fortress of *Ṯ3rw* (Zaru), see ERMAN's article *Ä. Z.* 43 (1906), 72—73. This was the starting-point of the great military road to Palestine followed by all the early armies; thus by that of Sethos I, cf. *p ḫtm n Ṯ3rw*, L., *D.* III 128b, completed by CHAMP., *Not. Descr.* II 94. Hitherto Zaru has been placed in the neighbourhood of the modern town Ismailiyeh; but Herr KÜTHMANN, whose researches on the subject will be published in a thesis entitled *Die Ostgrenze Aegyptens*, appears to have good reasons for placing it much farther to the North, in the vicinity of El Kantara.

3) *T'-t Ssw*, identical with *T'-t R'mssw-mry-'Imn*, which was reachable by boat from Zaru, see *Anast. V* 24, 8. Also doubtless the same as *T'-t p M3i* "The House of the Lion" in the Sethos reliefs, L., *D.* III 128a; this too is connected with Zaru by water, possibly, as KÜTHMANN suggests, the Pelusiac branch of the Nile.

4) *H(3)n(3)* "would that", cf. *Anast. IV* 11, 12; *Sall. III* 6, 7 (the hieroglyphic texts have here *h3*); cf. too *h3 my*, above 19, 6.

5) *Ḥ-ṯ-y-n* is very plausibly compared by MAX MÜLLER (*As. u. Eur.* 134) with the name of a well in L., *D.* III 128a (under the horse's tail), but all the copyists confirm the reading *ḥ-p-3-n*, (not *ḥ-ṯ-n*) there. The name is now destroyed, as N. de G. DAVIES, to whom I am deeply indebted for a collation of the Karnak reliefs, informs me.

6) Identical with *W3dy-t n Sty-Mr-n-Ptḥ* in L., *D.* III 128a. — *Mi rh*(read *nk?*)*r*, see p. 13*, n. 8.

7) In the Karnak scenes (L., *D.* III 127a) the next fortress to that of Buto is called *P-nḫtw n*(?) *Sty-Mr-n-Ptḥ*; with the natural change of the royal name Sethos into that of Rameses II (cf. last note), this name becomes clearly similar to *m py-f nḫtw Wsr-m3't-R'* in the papyrus. The meaning of *nḫtw* (masc. sing.) is not certain; possibly the preposition *m* should be omitted.

8) *S-b-ēl*, an unknown locality.

9) *'Ib-s-ḳ-b*, apparently a pool rather than a well, occurs in its right position L., *D.* III 127a.

10) The locality *'-y-n-n* is unknown.

11) *Tp-rd*, lit. "principle" or "rule", doubtless here means the "position" in relation to other places; cf. 28,8.

12) *N-ḫ-s* occurs in the Sethos reliefs, though it is not depicted in any of the publications. DAVIES writes to me that under the horses' tails in L., *D.* III 126b there should be inserted a small fortress over a pool or well; this bears the name "*N(3)-ḫ(3)-s(w)* (det. of water) of the Prince . . . . ."

13) Before *Ḥ-b-r-t* in the papyrus there is an *r*, which should probably be omitted. This name also possibly occurs, though in a damaged form, in the Sethos reliefs; it is the fortress at the top of L., *D.* III 126b, with the inscription "The town which his Majesty newly built at the well of *Ḥ-b*(?)-[*r*]-*t*" the stroke after the second letter shows that this can be neither *w* nor *3* as the publications give.

14) *R-pḥ*, Raphia, the modern Rafah, a town not far from the sea about 22 or 23 miles south of Gaza. Also mentioned in the palimpsest list of Sethos I, MAX MÜLLER, *Eg. Researches* I 57, 16; 58, 17.

15) *Wr*, see above p. 18*, n. 19. — The length of the *itr*, or schoenus, is not yet determined, see *Ä. Z.* 41 (1904), 58—60.

16) *Ḳ-d-t*, עזה, *Γαζα*, the southernmost of the Philistine cities; elsewhere in Egyptian spelt *G-d-t*, BURCH. no. 1071.

17) *Swh m* "to boast of", see *Admonitions* p. 28. — *M-r-y-n*, see above p. 25*, n. 1.

18) Hori goes on to contrast his own knowledge and skill with the ignorance of Amenemope. These sentences are very obscure in part.

19) *Ḫn-(r)-y*, see above p. 20*, n. 10.

brave man who can measure himself with me[1]! I am initiated in the decrees(?) of Month[2].

How marred is every (word) that cometh out over thy tongue! How feeble[3] are thy sentences! Thou comest to me wrapt up[4] in confusions, loaded with errors. Thou splittest words asunder, plunging ahead(?)[5]. Thou art not wearied of groping[6]. Be strong! Forwards! Get thee along(?)[7]! Thou dost not fall. What is it like not to know what one has reached[8]? And how will it 28,5 end[9]? I retreat[10]. Behold, I have arrived. Thy passion is soothed(??)[11], thy heart is calm. Do not be angry[12]. . . . .[13]. I curtail(?)[14] for thee the end of thy letter, I answer(?) for thee what thou hast said. Thy narratives are collected upon my tongue, established upon my lips. They are confusing to hear[15]; none who converses(?)[16] (with thee) can unravel them. They are like the talk of a man of the Delta with a man of Elephantine[17].

Nay, but thou art a scribe of the Great Gates, reporting the affairs of the lands, goodly and fair [to] him who sees it[18]. Say not that I have made thy name stink[19] before others(?). Behold, I have told thee the nature of the Maher; I have traversed for thee Retenu[20]. I have marshalled before[21] thee the foreign countries all at once, and the towns in their order. Attend(?)[22] to me, and look at them calmly[23]; (thus) thou shalt be found able to describe them[24], and shalt become a travelled(?) . . . .[25]

---

1) *Sṯn r ḥʿw-ỉ*, see above p. 15*, n. 10.

2) *Wn m* occurs once again on a writing-board in University College, London (= *Rec. de. Trav.* 19, 95), where *wn-ỉ tw* should be emended for *wn-twỉ*. The determinatives of *wḏ* are perhaps wrongly borrowed from *wḏb*. Month is here the war-god, so that the sentence is a further assertion of the martial qualities of Hori.

3) *Wỉ(ꜣ)wỉ(ꜣ)*, cf. *Berlin Ostracon* 10616; *Pap. Bibl. Nat.* 198, 2, 21; *Medinet Habu*, unpublished, under the Balcony in the southern colonnade.

4) *Bnd*, see p. 13*, n. 15.

5) Lit. "in entering before thyself"; the sense is apparently that Amenemope continues to write in haste, heedless of the injuries which his precipitate behaviour causes him to inflict upon the language. — *N ḥr-k*, see p. 25*, n. 2.

6) *Gmgm* "touchings" "fingerings"; for *gmgm* thus as the equivalent of ϭⲟⲙϭⲙ *palpare*, cf. *Harris 500*, recto 1, 2; 7, 12.

7) The writer sarcastically encourages Amenemope to persevere in his writing. — I suspect that the words *ỉmỉ ẖs-ꜣst-tw* are simply a corruption of *ms-tw* (cf. 15, 8), with elements borrowed from *ꜣs* "to hasten" and *sẖs* "to run".

8) I. e. thou art ignorant of thy plight.

9) *Pḥwỉ nn m ỉẖ*; this phrase occurs again on *Petrie Ostracon* 45.

10) Read *bhꜣ-ỉ*. Hori announces his intention of retiring from the contest.

11) *Hn-tw* might possibly be imperative: "give in!"

12) *Ḥḏn*, see above p. 10*, n. 16.

13) *Sy* is unknown; for *n ỉmw* cf. p. 27*, n. 10.

14) *Ḫʿḳ* lit. "shave", probably here in the sense "to curtail" or "to summarize".

15) Lit. "confused in hearing".

16) For *ꜣʿʿ* cf. *Israel stele* 22, where "conversing" clearly seems to be meant; so too perhaps *Sall. I* 8, 1. Compare too the difficult epithet (or title) *ꜣʿʿ* of which Spiegelberg has collected the examples *Rec. de Trav.* 14, 41.

17) This sentence is rightly often quoted in proof of the existence of dialects during the New Kingdom.

18) I. e. probably, "who sees what thou doest".

19) *Tm-k* optatively, see p. 11*, n. 4. — *Ḫnš*, Coptic ϣⲛⲟϣ, only here transitively.

20) *Tnw* is doubtless a corruption of *Rtnw*, as throughout in the great Berlin manuscript of *Sinuhe*.

21) Lit. "I have led to thee".

22) *Hn* "to bow" "incline", with the meaning "to attend to", cf. above 26, 3.

23) *Ḳb*, see above p. 9*, n. 7.

24) *Gm r*, see p. 17*, n. 3.

25) For the sense cf. 20, 6. Hori holds out to Amenemope the hope that he may some day appear to be a much-travelled warrior.

Appendix. The three technical problems of Anastasi I (14, 2—17, 2).

In the course of the controversy the scribe Hori propounds three problems connected with the building and erection of monuments such as a "royal scribe in command of the soldiers" might be called upon to solve. Amenemope's vaunted skill in his profession is thereby put to a very severe test, and in every instance he finds himself unable to reply. The technicalities of these passages are such that the modern Egyptologist is placed in a far worse quandary than this ancient scribe; so far from being able to supply the answers, he is barely able to understand the questions. I shall here attempt, as far as is possible, to define the nature of the three problems; in dealing with this difficult subject I have had the great advantage of consulting with Professor BORCHARDT, the first authority in such matters, and I am indebted to him for a number of valuable hints.

Problem 1 (14, 2—8). This deals with the building of a brick ramp of unusually large dimensions. The Egyptian word is *st3*, which etymologically means a place over which something is dragged or drawn. In the royal tombs the sloping, downward, passages were called *st3 ntr* "the divine passage". In *Piankhi* 91 *st3* is an ascending ramp used for scaling the walls of a hostile fortress. That *st3* here is an ascending ramp is clear from the description, the length being 730 cubits (more than 383 metres), and the breadth 55 cubits (nearly 29 metres). The ramp is said to consist "of 120 *r-g-t*", concerning which we later learn that each measured "30 cubits, by a breadth of 7 cubits". BORCHARDT conjectures with great probability that these were "compartments" ("Kästen") in the interior of the ramp, formed by brick partition-walls of no great thickness; these compartments would be filled with sand, a great saving of bricks thus being effected. A ramp constructed exactly in this manner has been found just to the South of the mortuary temple of the Second Pyramid, and belongs to about the 19th. Dynasty[1]. If the view of the *r-g-t* here taken be correct, the word may possibly be derived from *r3* "mouth" (cf. *r3-st3*, *r3-w3t*) and *g(3)t(3)* "shrine" "box" (BRUGSCH, *Wörterb.* 1520; *Suppl.* 1289; *Pap. Turin* 105, 21; 107, 19). With a length of 30 cubits, the *r-g-t* would leave a reasonable thickness of $\frac{55-30}{2} = 12 \cdot 5$ cubits for the exterior walls of the ramp[2]. On the other hand the indication that the "compartments" (*r-g-t*) were 7 cubits broad is impossible; this would already give $120 \times 7 = 840$ cubits for the length of the ramp, without reckoning either the thickness of the partition-walls between the compartments or that of the end-walls at the top and the bottom of the ramp. In spite of this serious difficulty, BORCHARDT's view of the *r-g-t* seems the only way of accounting for their number and their length.

The ramp, (i. e. its exterior walls) is stated to have been "filled with reeds and beams". This of course alludes to the practice of strengthening vast brick walls with reed-mats interposed between the courses and with transverse wooden beams inserted at a distance of some feet from one another. This mode of building is exemplified in the fortresses in the Second Cataract and elsewhere; see HÖLSCHER, *Das Hohe Tor von Medinet Habu*, p. 36.

The height of the ramp at its highest part was 60 cubits[3] and, if I understand the next words rightly, the height in the middle of its upward slant was 30 cubits. To me it

---

1) Regierungsbaumeister HÖLSCHER, in whose forthcoming book on the temple of the Second Pyramid a plan and section of the ramp will be found, tells me that the breadth of the compartments averages about 3 · 5 metres, i. e. approximately 7 cubits.

2) Measured at the top; the thickness at the bottom would be 27 · 5 cubits, owing to the batter, if my hypothesis (see below) be correct.

3) This gives a slant of 8 · 2 cubits in every 100; that of the ramp leading to the pyramid of Ne-user-Re was equal to 7, 75 cubits in every 100 (BORCHARDT, *Das Grabdenkmal des Ne-user-re*, p. 44).

extent of the divergence from the perpendicular, which amounted to 1 cubit and 1 finger(?) on the entire height of 110 cubits, or 0·26 fingers in every cubit[1]. The pyramidion (*ty-f brbr*) is stated to have been 1 cubit in height, of course an impossibly small measurement. Lastly, the point of the pyramidion (*ḥwy*, only here) is said to have measured two fingers; BORCHARDT understands this to refer to the length of the sides of the tiny square surface at the summit of the pyramidion. For all these measurements, see the figure on the preceding page. All the dimensions required for determining the content of the obelisk, and hence also its weight, appear to be given.

Problem 3 (16,6—17,2). The last problem is, of the three, by far the most difficult to understand. It is at all events clear that it concerns the erection of a colossal statue, and that this statue had to be gradually lowered to its ultimate position by the removal of a great artificial magazine of sand on to which it had been hauled(?). The statue covers 30 cubits as it lies stretched upon the ground, and has a breadth of 20 cubits. The word for magazine is *mḫr*, the ordinary word for a store-house in which corn or other things were kept; and its sub-divisions(?) are called *šmm*, also a known word. All the other technical terms and dimensions are quite obscure. The *šmm* are stated to have been filled with sand from the river-banks, and the magazine, which was situated under (*ẖr*) the colossus, has to be "emptied" (*sšw*). The question asked of Amenemope is as to the number of men to be employed in order to remove the sand (*ḫm* "demolish" "overturn") in six hours. The use of sand for the gradual lowering of monuments is, as BORCHARDT points out, exemplied by the late tombs near the Pyramid of Onnos (see BARSANTI's article, *Annales du Service* I 283—4) where the massive lid of a sarcophagus was supported upon wooden pillars resting upon sand; when it was desired to bring the lid into its final position, the sand was gradually allowed to escape from under the wooden pillars[2].

1) It is probable that the *ḥ* after *mḥ* 1 should be emended into *ḏbʿ*. The batter of the obelisk is quite unusually steep.

2) One may also compare BONOMI's theory of the manner in which the colossus known as the Vocal Memnon was erected (*Ä. Z.* 45 [1908], 32—34); however BORCHARDT is of the opinion that sand cannot have been used in that case.

II. A collection of model letters.

Description of the manuscript. The Papyrus Koller (*Pap. Berlin* 3043) is known to have passed directly from the collection of Baron KOLLER into that of the Berlin Museum, but no record has been kept of the exact date when this occurred. The manuscript measures 136 cm in length, and has the normal height of 21 cm. It is on the whole well preserved, though it has suffered damage through rough handling. A Museum register dating from the days of PASSALACQUA informs us that the verso once exhibited a drawing of the lower part of the double crown of the Pharaohs, and that there also was a brief hieratic inscription; but the manuscript having been early gummed upon cardboard, this statement can no longer be checked. At a subsequent period the edges of the cardboard around the lacunae were carelessly trimmed, with disastrous results. At present the papyrus is safely preserved under glass in four sections. The writing of the *recto* is upon the horizontal fibres, and comprises five complete columns or pages; these consist of eight lines apiece, except the second page, which has nine lines. It is well-nigh certain that some pages are missing at the end, since the text breaks off in the middle of a sentence. Whether anything is lost at the beginning is more doubtful, the main text of the first letter being complete, though the salutations that usually precede are absent. The handwriting is of the fine type characteristic of Ramesside literary papyri, and may be dated approximately to the end of the 19th Dynasty. If the criteria set up by MÖLLER[1] can be trusted, the papyrus is a product of the Memphite school of calligraphy. There are no verse-points, but the usual sign, written in red, serves to divide the sections from one another. A. WIEDEMANN printed a not entirely satisfactory hand-facsimile of the manuscript in his book *Hieratische Texte* (Leipzig 1879), Tafel X—XIV, and up to the present this has remained the sole edition of the text.

Contents and general remarks. Many of the finest literary papyri of the New Kingdom belong to a class to which ERMAN has aptly given the name of *Schülerhandschriften*[2]. They are the work of youthful scribes employed in one or other of the public administrative departments, where they seem to have received, after the manner of apprentices, some tuition from the superior scribes. Often the appointed task was the copying of some wellknown literary piece, such as the *Instructions of Amenemhet I to his son*, or the *Poetical account of Rameses II's victories;* in such cases it is impossible to detect the pupil's hand, unless it betrays itself by dates jotted down in the margin to indicate the amount written daily, or by the presence of corrections in the teacher's hand[3]. Often however the subject-

---

1) *Hieratische Paläographie* II p. 2—3.

2) On the whole subject of the *Schulhefte* or *Schülerhandschriften*, see ERMAN, *Aegypten* 446—448.

3) Purely calligraphical corrections, such as that at the top of p. 3 of *Anastasi I*, seem sufficient proof of a *Schülerhandschrift*. — For dates see ERMAN, *loc. cit.*; they occur very frequently both in papyri and on ostraca.

matter of the texts copied ill conceals their educational aim; this is particularly the case with the collections of model letters, or miscellanies [1], of which the *Papyrus Koller* affords a very typical example. The letters contained in these miscellanies are of diverse kinds. The simplest consist of little beyond the elaborate salutations demanded by Egyptian good-breeding. Others, even less readable, are mere lists of articles to be manufactured or foodstuffs to be provided and are simply designed for the purpose of widening the pupil's vocabulary. The majority concern such commissions or affairs of everyday life as might later claim the scribe's attention in the course of his professional career. When the subject-matter permitted, the pupil frequently substituted his own and his teacher's names for those of the original writer and recipient of the letter; *Koller* 5,5 for example, mentions the names of two scribes, of which Amenemope may be that of the master, and Paibēs that of the pupil [2]. Besides letters, the miscellanies here described often contain short compositions of a more purely literary character; hymns to Thoth or Amon, eulogies of the Pharaoh or of the Capital, and above all homilies (as a rule not lacking in humorous touches) on the dignity of the scribe's profession.

The *Koller* comprises four letters, the first lacking the customary salutations and the last ending abruptly after a few opening sentences. The subjects are as follows:

a) The equipment of a Syrian expedition (1,1—2,2).
b) Warnings to an idle scribe (2,2—3,3).
c) A letter concerning Nubian tribute (3,3—5,4).
d) An order to make preparations for Pharaoh's arrival (5,5—5,8).

Short introductions with bibliographical notes being prefixed to the translations of each section, only a few general remarks are here necessary. The *Koller* is particularly closely related to the London papyrus *Anastasi IV*, with which it has two sections in common. The orthography is good, and mistakes or corruptions seem to be relatively few. The pupil by whom the *Koller* was written seems to have aspired to erudition, for the texts chosen are full of technical and foreign words.

The first to translate the papyrus was A. WIEDEMANN, whose renderings (*op. cit.* p. 19—23) are accompanied by brief notes on the subject-matter but without a philological commentary. The only other treatment of the papyrus as a whole is that of ERMAN in the handbook entitled *Aus den Papyrus der königlichen Museen*, Berlin 1899, belonging to the official series of *Handbücher der königlichen Museen zu Berlin* (p. 93—97).

a. The equipment of a Syrian expedition.

This is a short model letter describing the preparations to be made for an expedition to Syria and enumerating in detail the horses, attendants, chariots and weapons that have to be made ready. The point of the composition doubtless lay in its copious use of foreign and technical words, which would serve at once to exhibit the teacher's erudition and to increase the store of the pupil's learning. No duplicate of this letter is known, but it is shown by its last words to be closely related to a letter preserved entire in *Anastasi IV* (13,8—end), the beginning of which is found on the last page of the *Koller*. Other texts which mention the various parts of the chariot and the weapons contained in it are

---

1) The best-known of these are *Anastasi II. III. IV. V*; *Sallier I*, in the British Museum; *Pap. Bologna 1094*; *Pap. Leiden 348*, recto.

2) It is not always the name of the pupil which stands first, as may be seen by comparing *Anast. III* 1, 11 with *ibid.* 3, 9. It is curious that the names of the scribes in *Anast. III* are Amenemope and Paibes as in *Koller*; the papyri do not seem to be written by the same hand.

Anastasi I and the *Edinburgh Poem about the Chariot* (*Ä. Z.* 18 [1880], 94—95). No translations of this section seem to have published besides those named above in the general introduction.

⟨The scribe Amenope writes to the scribe Paibēs[1],⟩ saying: — Take good heed to make ready the array(?)[2] of horses which is (bound) for Syria, together with their stable-men[3], and likewise their grooms[4]; their coats[5] . . . . -ed and filled with provender and straw, rubbed down twice over; their corn-bags(?)[6] filled with kyllestis-bread[7], a single ass(?) in the charge of[8] (every) two men. Their chariots are of *bry*-wood(?)[9] filled with ⟨all kinds of(?)⟩ weapons of warfare[10]; eighty arrows in the 1,5 quiver[11], the . . . .[12], the lance(?)[13], the sword[14], the dagger, the . . . .[15], the . . . .[16], the whip[17] of *ṯ3g*-wood[18] furnished with lashes[19], the chariot-club[20], the staff(?)[21] of watchfulness, the javelin[22] of Kheta, the rein-looser(?)[23], their facings ⟨of⟩ bronze of six-fold alloy[24], graven with chiselling(?)[25], ... -ed, and . . . . . . -ed[26]. Their cuirasses[27] are placed beside them. The bows are adjust-

1) The names are restored from 5, 5.

2) *Rḳs* is possibly identical with the rare Hebrew collective word for "horses" רכש; see BURCH. no. 642, where the phonetic difficulty is pointed out. — For *grg* we should expect *grg-tw* (cf. 5,6), but in 3, 5 and *d'Orbiney* 2, 2 the ending is similarly omitted.

3) *Ḥry iḥ*, a very common title, which in itself probably indicates quite a low rank.

4) *Mri*, see p. 6*, n. 16.

5) *Šnw* lit. "hair", only here of horses' coats.

6) Here the provisions for the stable-men and grooms appear to be referred to; these were carried on asses. — *Ḫ3r*, properly a corn-measure, is occasionally determined with the sign for the hide, cf. *Rhind Math. Pap.* 41, 3. 4; 43, 1, MAR., *Karnak* 54, 46.

7) The *k-r-š-t* was a small loaf weighing from about half to three-quarters of a pound, see EISENLOHR, *Proc. S. B. A.* 19, 263; the name is preserved in the Greek κυλλῆστις, a word known from Hdt. II 77 and other sources (see *Ä. Z.* 47 [1910], 159 footnote).

8) *R iwd* as preposition means either (1) "in the charge of", so here and *Pap. Bologna* 1094, 6,7; *Pap. Bibl. Nat.* 187, 3, 4. 6; or (2) "with" (*apud*) cf. *Salt 124*, verso 1,1; *Pap. Turin* 57, 1; 103, 1, 16.

9) *Bry*, only here and *Anast. IV* 16,7, where it is likewise a kind of wood of which chariots were made.

10) Probably emend *ḫꜥw ⟨nb⟩ n rꜥ ḫ-t*; note that in this expression *ḫ-t* is written with *t* and stroke during the 18th. Dynasty, (cf. *Urk.* IV 699).

11) *'Is-p-t*, see above p. 27*, n. 17.

12) *Ḥmy-t*, as a weapon belonging to the chariot, *Edinburgh Poem about Chariot*, recto 11.

13) *M-r-ḥ*, only here and *Anast. IV* 17,1; cf. ⲙⲉⲣⲉϩ, but this comparison perhaps fails if the Boheiric form ⲙⲉⲣⲉϧ quoted by Peyron is well authenticated.

14) *Ḥ-r-p*, Hebrew חֶרֶב, also *Anast. IV* 17,1; *Edinburgh Poem*, recto 13.

15) *Ḳ-w-t*, only here.

16) *Sk-hm(w)*, again only *Anast.* IV 17, 1.

17) *'Is-b-r*, see BURCH. no. 134 and above p. 26*, n. 6.

18) *Ṯ3g*, a species of wood; whips are made of it, as here, *Anast. IV* 17, 2; the chariot-pole, below 2, 1; chariots, *Urk. IV* 707; the word also *ib.* 701. 705. 732.

19) *Rwd-(wt)* "lashes", only here; *Anast. I* 26,8 uses another word; for *rwd-(wt)* as bow-string cf. below 1,8; LACAU, *Sarcophages II*, Index.

20) So too *t ꜥwn-t n ty-k mrkb-t, Edinburgh Poem*, verso 9.

21) *Ḥꜥ(w)*, elsewhere only *d'Orbiney* 13,1 (conclusive as to sense); *Anast. IV* 17,3.

22) The weapon *niw* seems from the hieroglyphic determinative in R., *I. H.* 215, 31; 240, 37; 241, 44 to be a javelin; for javelins, see WILKINSON, *Anc. Egyptians* (ed. BIRCH), I 208. Cf. *p niw n ty-k mrkb-t, Edinburgh Poem*, recto 11.

23) For *ṯt* "to loose" see p. 22*, n. 7, and for *ḫnr* see p. 20*, n. 10; what instrument is here meant is hard to say.

24) *Sm(3) n sis* is evidently an alloy of six ingredients; from this passage and from *Harris I* 45, 5; 47, 6; 52 b 9 it is plainly a kind of bronze, and from *ibid.* 6, 9; 47, 4 we know that it was of the colour of gold, i. e. probably like brass. Other references, *Harris I* 59,3; *Pap. Turin* 32,7. 9; *Anast. IV* 16,12.

25) For this difficult phrase see p. 28*, n. 13.

26) *Fti* and *m-s-ḳ*, unknown verbs.

27) *R-b-š-y*, from Hebrew לבש; see SPIEGELBERG, *Petubastis*, Index no. 235—236 for demotic instances and some important remarks.

2,1 ed(??) to their strings[1], their wood[2] being tested in drawing, their(?) bindings(?)[3] consisting of clean .... leather(?)[4]. The pole[5] is of *ṯ3g*-wood[6], ... -ed[7], shaped(?)[8], fitted with leather[9], finished off(?), oiled[10] and polished(?)[11].

1) *Rwd-(wt)*, see above note 18; the meaning of *ṯsy* here is obscure.

2) *Dbw* lit. "horns", here clearly the wooden part of the bows; the word for bow in other languages not seldom alludes to its shape, cf. *arcus*, *Bogen*.

3) *M-š-y*, only here and *Anast. I* 26,6.

4) For *mtr-t* we may possibly compare *Pap. Kahun* 19, 57.

5) For the word ʿ see p. 28*, n. 6.

6) *Ṯ3g*, see p. 37*, n. 18.

7) *G-p* does not occur in the parallel text *Anast. IV* 16, 11—12, and has evidently nothing to do with *g-p* below 2,8; it may be for *g-r-p*, a gloss on *g-r-b* (see next note).

8) *G-r-b* (so too *Anast. IV* 16, 11) is identical with *g-r-p*, *Anast. I* 26,5; see p. 28*, n. 7.

9) *Tby* is clearly derived from *ṯbw* "sandal-maker", "leather-worker"; in the sense "shod", cf. *Anast. III* 8, 6 = *Anast. IV* 16, 5.

10) *Sgnn* "to oil", cf. *Anast. III* 8,4 = *Anast. IV* 16,3; *Anast. IV* 15,4; 16,12; *Sall. I* 4,10; 5,3; *Harris 500*, verso 5,9.

11) *M-š-r-r* (also in *Anast. IV* 16,12) looks like a Semitic passive participle **mashrūr;* the word is unknown.

b. Warnings to the idle scribe.

Almost all the great miscellanies of the New Kingdom contain threats and warnings addressed to the idle scribe, most of which begin with the stereotyped words found here (e. g. *Sall. I* 6,1; *Anast. IV* 11,8; *Anast. V* 6,1). The present text, fragmentary duplicates of which are found in *Anastasi IV* 2,4—3,2 and *Anastasi V* 5,1, is peculiar in the fact that it consists almost entirely of a long drawn out simile, the pupil being compared to a careless sailor. The end of the section is much damaged and practically unintelligible. The parallel texts from *Anast. IV* and *Anast. V* are reproduced in facsimile in the *Select Papyri in the Hieratic Character* (London 1842 and 1844), plates LXXXIII. LXXXIV and plate XCIX respectively; for a complete description of these Mss. the reader must be referred to a subsequent instalment of this work. A few phrases were translated by CHABAS in his *Voyage d'un Égyptien*, pp. 141. 241, and the whole, so far as it is preserved in *Anast. IV* and *Anast. V*, was rendered into French by MASPERO, *Du genre épistolaire chez les anciens Egyptiens* (Paris 1872), p. 28—30. The version of the *Koller* was first utilized by LAUTH, *Die altägyptische Hochschule von Chennu*, in *Sitzb. d. k. Bayr. Akad. d. Wiss.*, 1872, p. 66; then more completely by WIEDEMANN in 1879. The only other translations seem to be those of ERMAN in *Aegypten* (1885), p. 514 (mainly a paraphrase) and in the handbook mentioned above in the general introduction.

They[1] tell me that thou forsakest writing[2], and departest and dost flee; that thou forsakest writing and usest thy legs[3] like horses of the riding-school(??)[4]. Thy heart is fluttered; thou art like an *ʿḫy*-bird[5]. Thy ear is deaf(?)[6]; thou

1) The section opens with the usual epistolary formula *r ntt* "to wit", which is best omitted in translating.

2) Not "books"; for *sš* "to write" so determined cf. *Anast. V* 8,3; *Inscr. of Mes* N 14.

3) The expression *mḥ-k m rdwi-k*, lit. "thou seizest (or "art full of") thy legs", seems to occur only here; *Anast. IV* has a superfluous *m* before *mḥ*.

4) *Ṯ-ḥ-b* only here; the sense is quite problematical. — *Ḥtri* seems never to be used in late-Egyptian for simply one horse; it means a pair, or a horse and chariot.

5) The bird *ʿḫy* is also mentioned *Anast. IV* 1b, 1; *Pap. med. Berlin* 21,2.

6) *D-n(r)-g* only here; however the proper name *D-n(r)-g* (BURCH. no. 1189), older *D3g* (e. g. *Cairo stele M. K.* 20007; L. *D.* II 147b), is often determined with the ear, implying that the verb was common.

art like an ass in taking beatings[1]. Thou art like an antilope in fleeing. Thou 2,5 art not[2] a hunter[3] of the desert, nor a Mazoi[4] of the West! Thou art one who is deaf and does not hear[5], to whom men make (signs) with the hand. Thou art like the mate of a skipper skilled in (managing) the boat. When he is skipper[6] in the boat, he stands at the prow(?)[7], he does not look out for dangerous winds, he does not search for the current[8]; if the outer(?)[9] rope is let go, the rope in front(?)[10] is in his neck(?)[11]. When he is pulling the rope[12], he catches(?)[13] the . . . . . . -birds, he plucks[14] . . . . . -flowers(?)[15] on the banks, he cuts away[16] clods of earth(??)[17]. His . . . . . .[18] . . . . -trees, he . . . . *ksb*-trees[19]. His . . . . . .[20] is of seven cubits, he cuts reeds(?). His tresses(??)[21] 3,1 . . . . . . to his feet, in work of Kush. His . . . . is of bright . . . .[22] in work of the overseer of . . . .[23]. He binds threads[24] to its(??) end, in

1) I. e. thou art as stubborn and unresponsive when punished as a beaten ass.

2) *Anast. IV* wrongly omits the negation here and inserts it before *mnth sḥy*.

3) For similar writings of *nw* "hunter" (for the sense see the determinative in *Urk. I* 2) cf. *Urk. IV* 994; *Harris I* 28,4; L., *D.* III 356.

4) For the Mazoi as hunters, cf. especially *Anast. IV* 10,5; in *Urk.* IV 994 the titles "great one of the *Mḏ3w*" and "overseer of hunters" occur in parallelism with one another. See now Max Müller, *Egypt. Researches* II p. 76—77.

5) Cf. *Anast. I* 6,6; 26,3.

6) I. e. when the mate (*ḥrỉ-ʿ* "subordinate") is acting as the skipper or *rêis*.

7) *Tpti*, probably a special nautical term; *tp-t* is the name of a part or an appurtenance of a boat made of cedar-wood and measuring 20—30 cubits (*Turin. Pap.* unpublished); *tpti* may be a *nisbe*-form from it.

8) *H-(y)-n*, Coptic ϩⲟⲉⲓⲙ (*Rec. de Trav.* 28,214); elsewhere the word clearly means "waves", not "current", e. g. *Anast. IV* 1b, 2.

9) If *n bn(r)* is read with *Koller*, this must be a genitive qualifying *p nwḥ*; if the reading of *Anast. IV* be preferred *ḫ3ʿ r bn(r)* means "to let go".

10) *Ḫnt-t* is possibly a technical term for the rope attached at the prow.

11) I. e. probably, becomes entangled with his neck. Wiedemann takes this clause to mean, "the towing-rope is placed round his neck", but this cannot be correct, as the sentence clearly continues the description of what happens when the mate is acting as skipper. — *Ḫḥwi-k*, which Koller reads instead of *ḫḥwi-f* (*Anast. IV*), is certainly a mistake, like *rdwi-k* in 3,1; the scribe is still influenced by the pronouns of the second person with which the text began.

12) *Wnn-f ỉtḥ p nwḥ* (parallel to *wnn-f m nfw* above) introduces a new picture of the mate's heedlessness; he is now imagined as towing on the river-bank. For *ỉtḥ p nwḥ* the Modern Egyptian would say quite similarly يجرّ لبان.

13) *G-p* is here probably the equivalent of *kp* (see above p. 9*, n. 14); the substitution of *g* for *k* is however difficult to parallel at this period.

14) *Fg*, so written also *Pap. Leiden 345*, recto G 2,1, is the equivalent of *fḳ* in *Pap. Turin* 89,5; doubtless Coptic ϥⲱϭⲉ: ϥⲱϫⲓ *evellere*.

15) *Šrỉ(w)*, only here.

16) *Pns* means "to cut off" the head (e. g. Mar., *Mast. D* 10; *Mission V* 622) or the ribs (cf. Nav., *Deir el Bahari* 107; *Mission* V 617) of a bull to be sacrificed; also "to cut away" hair, *Ebers* 63,13. Here I imagine the meaning to be that the idle sailor, instead of attending to his towing, amuses himself with hunting the birds or picking the flowers on the bank, or else in knocking away the heavy lumps of muddy earth at the edge of which he is walking.

17) *M-ḳ-ḳ*, cf. *Anast. III* 2,4, where the word occurs together with *3ḥ-t* "fields".

18) *Mtr-?* here is of unknown meaning.

19) Before *ksb*, a kind of tree mentioned already in the Pyramidtexts (e. g. 456. 994), an infinitive is expected; perhaps emend *ỉn* "to bring".

20) *W3ḥ-t*, meaning unknown.

21) *Nbd-t* means "tressed hair" in *d'Orbiney* 10,7. 9; 11,2. 4; *Anast. III* 3,3; masculine in *Harris 500*, recto 6,1. The damaged word following is perhaps a verb "to dangle" "hang down" or the like. For the erroneous reading *rdwi-k* of Koller see above n. 11. — The very obscure sentences which follow may allude to the foppish attire of the skipper's mate, another sign of his disinclination for regular work.

22) The words *ỉ-t-t* and *n-f-ʿ* are ἅπαξ λεγόμενα.

23) Reading and meaning uncertain; the title occurs again *Harris I* 7,9 (see *Ä. Z.* 23 [1885], 60—61), where its connection with cattle suggests that it means "overseer of horns".

24) *G-w-tn* seems from *Anast. I* 24,6 to mean "to bind" or "tie". — *P-t-r* probably the Hebrew פתיל, see Burch. no. 430. The meaning of the sentence is quite obscure.

order to wear a loin-cloth(??)[1]. He is one who pricks up(?) the ear[2] on the day of the ass; (he is) a rudder on the day of the boat. I will do all these things to him[3], if he turns his back[4] on his office.

---

1) This garment, the transcription of which is unknown, is often mentioned in late-Egyptian texts e. g. below 4,6; BERGMANN, *Hierat. Texte* 1,2. 5; *Pap. Berlin* 9784,6; *Harris 500*, verso 1,11; and six times in *Harris I*.

2) *Mḥ msḏr*, lit. "to fill the ear", i. e. probably to "listen" "hear" "hearken"; cf. *Anast. I* 20,6 (with a superfluous *m*); *Anast. IV* 5,4; both rather obscure passages.

3) These words must refer to the lazy pupil addressed in the first words of this effusion, and we should therefore expect the second person; this however is given by neither text, though *Koller* has had two examples of the suffix 2nd. pers. sing. where it was not required (see p. 39*, n. 11). Nor is it plain to what the words "all these things" refer, since no punishments have been threatened or described.

4) *Ḫ3ʿ ḥ3-f r*, cf. *Anast. V* 6,1; 15,7 = *Sall. I* 6,2; *Pap. Turin* 88,11; *Israel stele* 11; *Berlin Ostracon* 11247 (*Hierat. Pap.* III, 35).

c. A letter concerning Nubian tribute.

This is a letter supposed to be sent by a high official named Paser, probably the Viceroy himself, to a Nubian chieftain, ordering him to make ready the tribute of his district without delay. The interest of the section centres in the long enumeration of the Nubian products, the best list of the kind that exists among our literary records. A shorter but interesting list occurs in the letter from Rameses XII to his Viceroy Pinehasi (PLEYTE-ROSSI, *Papyrus de Turin* 66. 67). A duplicate of the beginning of the present letter, with a rather different text, occurs on a potsherd discovered by QUIBELL in the course of his excavations at the Ramesseum; it has been published by SPIEGELBERG in his volume *Hieratic Ostraka and Papyri* (Egyptian Research Account, extra volume, 1898), plate 5, no. 39a. Besides the translations by WIEDEMANN and ERMAN already mentioned, there is an incomplete rendering by H. BRUGSCH in his book *Sieben Jahre der Hungersnoth* (Leipzig 1891), p. 115.

The fan-bearer at the right hand of the king, the captain of auxiliary troops, the overseer of the countries of Kush, Paser[1], writes to him who protects 3,5 his people[2], to wit: — This communication is brought to thee saying: when my letter reaches thee thou shalt cause the tribute to be made ready[3] in all its items[4], in *ỉw3*-bulls, young *g3*-bulls, *wndw*-bulls, gazelles, oryxes[5], ibexes[6], ostriches[7]; their broad-boats, cattle-boats[8] and (ordinary) boats being ready to hand(?)[9], their

---

1) The titles are unusual, but "fanbearer" etc. and "overseer of the lands of Kush" suggest that the "Royal Son of Kush" is meant. One "Royal Son" or "Viceroy" named Paser lived in the reign of Eye (stelae at Gebel Addeh); another seems to be mentioned under Rameses II on a monument at Naples, see BRUGSCH, *Thesaurus*, 953.

2) *Mỉk rmṯ-f* is found nowhere else either as a title or as a proper name. SP., *H. O.* had a different reading: "[to .... the overseer(?) of the ...... of] Kush".

3) Variant of SP., *H. O.*, "thou shalt take heed to have [the tribute] made ready".

4) Lit. "in all its things".

5) *M3-ḥḏ*, for this animal see VON BISSING, *Mast. d. Gemnikai I* 34, and for the spelling cf. *Harris I* 4,8; 20a, 12. 13 etc.

6) *Nr3w*, a late writing of the old word *nỉ3w*, found first in *Benihasan* II 4. 13; for the animal, see VON BISSING, *op. cit.*, p. 35.

7) *Nỉw*, already *Pyr.* 469a; a good instance *Urk.* IV 19; cf. too below 4.1.

8) *Ḫn-ỉḥw*, cf. *Anast. IV* 6,11; 7,6; *Harris I* 12b, 11; 69,13.

9) *Tpỉ-dt*, only here.

skippers and their crews prepared for starting; much gold wrought into dishes[1], refined gold(?) in bushels(?), good gold, precious stones(?)[2] of the desert in bags 4,1 of red cloth, ivory and ebony, ostrich[3] feathers, nebḳ fruit in . . . . , bread of the nebḳ[4], *š-ḳ-r-ḳ-b-y* of(?) *m-y-n-y-ḥ-s*[5], *ḥ-ḳ-ḳ*[6], *šs-y*[7], panther skins, gum[8], *dìdy*-berries[9], red jasper[10], amethyst(?)[11], crystal[12], cats of *Miw*[13], baboons, apes, *šnw*-vessels containing *ḥnti*(?)-pigment[14], cyperus-roots(?)[15], sacks(?) and *ps*-packets(??)[16]; numerous men of *'Ir-m-ì*[17] in front of the revenues, their staves(??)[18] adorned with 4,5 gold, ...[19] containing . . . . ,[20] . . . . -ed[21] with . . . .[22]

---

1) *Dd-t*, so written too *Anast. IV* 16,2, is a flat dish, cf. *Urk.* IV 631; the word occurs already in the Old Kingdom, cf. L., *D.* II 28, and its connection with the Hebrew דוּד seems very doubtful.

2) *Gmw*, an unknown word.

3) *Nrw* is a spelling for *nìw* (see above p. 40*, n. 7) cf. *Pap. Turin* 125, 6.

4) The tree called *nbs* in Egyptian, in Old Coptic ⲛⲟⲩⲃⲥ (*Ä. Z.* 38 [1900], 87), is the zizyphus, or Christ's thorn-tree, called by the Arabs نبق; its fruit is sweet and palatable when ripe; the "bread of the *nbs*-tree", mentioned here and in the lists of offerings, is probably a cake made from the dried fruit. See Maspero's article *Proc. S. B. A.* 13, 496—498. — *Ḥmḥm* is unknown.

5) Two unknown Nubian words, of which the first is probably the name of a fruit, the second that of a tree.

6) *Ḥ-ḳ-ḳ*, a fruit of some kind, conjectured by Brugsch to be that of the dum-palm, see *Wörterb. Suppl.* 855; further examples in the *Petrie Ostraca* 31. 37; for the spelling here cf. *Harris I* 19b, 12; 36b, 5.

7) *Šs-y*, probably a fruit, is mentioned next to *ḥ-ḳ-ḳ* in *Harris I* 65a, 4. 5 and again *ibid.* 74, 3. 4.

8) *Ḳmy* "gum", κομμι in Greek and ⲕⲟⲙⲏ in Boheiric, see especially Krall, *Studien z. Gesch. d. alten Aegypten* IV, 27—30.

9) *Dìdy*, according to Brugsch (*Ä. Z.* 29, [1891], 31—33) the magically potent red berry of the mandrake; this plant was particularly abundant in Elephantine, cf. besides Brugsch's references *Urk.* IV, 55. Pliny (*Hist. Nat.* 24, 102) mentions a plant called *ophiusa* which grew at Elephantine and possessed very remarkable properties.

10) *Ḥnm-t*, probably either red jasper or carnelian; for the colour cf. L., *D.* III 117 = Champ., *Not. Descr.* I 479; *Zauberspr. f. Mutter u. Kind* 1,1; *Turin Love-songs* 2,3. The knot-amulet, which is usually of jasper or cornelian, is said in *Totb. ed.* Leps., 159 to be made of *ḥnm-t*. The Hebrew אחלמה was compared with *ḥnm-t* by Brugsch, *Wörterb.* 1100; but LXX interprets this as ἀμέθυστος. *Ḥnm-t* as a product of Nubia, cf. *Pap. Turin* 67, 11; *Sehel, Famine Inscription* 16.

11) *Ḥ-m-ḳ* is without doubt merely a variant writing of *ḥm3g3-t*, which is mentioned elsewhere as a product of Nubia, see Brugsch, *Sieben Jahre der Hungersnot*, p. 129. Lepsius-Berend (*Les Métaux*, p. 21 footnote) cites Hoskins for the fact that this precious stone is depicted as red. If this evidence be relied upon, *ḥm3g3-t* might be carnelian, *ḥnm-t* being red jasper; or *vice versa*. However a word for "amethyst" is still wanting, and I prefer provisionally to identify *ḥm3g3-t* with this.

12) *'I-r-ḳ-b-s* is rightly identified by Bondi, *Lehnwörter* 28, with Hebrew אלגביש Assyrian *algamišu*.

13) *Miw*, a Nubian district, mentioned *Urk.* IV 796 and often.

14) *Ḥnti*(?) is often mentioned as a material used in writing, e. g. *Pap. Leiden* 347, 12,9; *Totb. ed.* Nav., 130,44; 134,17; together with other colours used for writing or painting, cf. *Mission* XV 26 (Luxor); *Ostracon Cairo* 25247; *šny*-vessels containing *ḥnti*(?), cf. *Harris I* 65a, 3; 74,2. Possibly this is the name of the raw material out of which the ordinary red paint was made.

15) *Nfw* is compared by Newberry (*Proc. S. B. A.* 22, 146—148) to a vegetable product, the root of the *Cyperus esculentus*, L., still known in the Sudan under the name نفو.

16) *Ps*, cf. *Harris I* 65a, 8; 74,6, in connection with *nfw*.

17) *'Ir-mì*, a well known tribe of the Sudan, cf. *Urk.* IV 333. 796; L., *D.* III 218c and elsewhere. Maspero (*Rec. de Trav.* 8,84) recalled the fact that the Galla race call themselves *Ilm Orma* "the sons of Orma", and concluded that the ancient word *'Ir-m-ì* is preserved in the modern *Ilm*; Tomkins (*Rec. de Trav.* 10,98) prefers to identify *'Ir-m-ì* with Orma, and if the name *Ilm Orma* has anything at all to do with *'Ir-m-ì*, this surely is the more likely view of the two, *Ilm* meaning simply "sons of" like Arabic بنو.

18) *'I-b-r-d*(?)*-ì*, an unknown word.

19) *K-r-k-r-t-b-ì* is unknown, and the verb(?) *s-š3-w* has a very suspicious appearance.

20) *S-n(r)-w* is a ἅπαξ λεγόμενον.

21) *Tftf* in *Pap. Turin* 74,5 means "to be disturbed" "fluttered" like the simplex *tfì*; its sense here is unknown.

22) *Ḥpì-t*, another unknown word.

and with all precious stones; tall men of *T-rk*[1] in loin-cloths(?), their fans[2] of gold, wearing high feathers[3], their bracelets(??)[4] of woven thread; many Negroes of all sorts[5]. Increase thy contribution every year! Have a care for thy head, and turn thee from thy indolence. Thou art [old]; look to it diligently[6], and beware! Be mindful of the day[7] when the revenues are brought, and thou passest 5,1 into the Presence beneath the Balcony[8]; the nobles ranged on either side in front of his Majesty, the chiefs and envoys[9] of every land standing gazing and looking at the revenues. Thou art afraid and shrinkest(?)[10], thy hand grows feeble, and thou knowest not whether it be death or life that lies before thee[11]. Thou art profuse in prayers [to] thy gods: "Save me, prosper me[12] this one time!"

---

1) *T-r-k* does not seem to occur elsewhere, unless the name hitherto read *ꜣ-r-k* (Urk. IV 796; de Morgan, *Cat. d. Mon.* I 67) should really be read with the *tiw*-bird.

2) *Bh-t*, cf. *Harris I* 21a, 2; *Harris 500*, recto 3,11; the phrase *ḥbs bh-t* "to carry(?) the fan" cf. Davies, *Amarna* VI 20; *Anast. III* 8,6 = *Anast. IV* 16,5; L., *D.* III 218c.

3) Erman regards the words "high of feathers" as an attribute of the fans; this appears to me less probable.

4) *K-r-m-t* occurs again only in the description of Negroes *Anast.* III 8,7 = *Anast. IV* 16,6 "their *k-r-m-t* are upon (*r*) their hands".

5) *M tnw nb*, see above p. 7*, n. 6.

6) For *bʿ*, see Erman's remarks *Ä. Z.* 42 (1905), 107.

7) Cf. *sḫꜣ nk hrw n ḳrs*, *Sinuhe B* 190.

8) *Sšd*, see p. 17*, n. 6.

9) *Mšʿyti*, only here; from *mšʿ* "to travel".

10) *Tḫbḫ* is probably to be emended into *tnbḫ*, for which see *Admonitions* p. 67.

11) For *n . . . n* "whether .... or", see *Inscription of Mes*, p. 16.

12) *Swḏꜣ-i*, probably infinitive used as imperative, see Sethe, *Verbum* II § 566; see above p. 22*, n. 17, and cf. *in-i*, *Pap. Bibl. Nat.* 197, 4, 6; *int-i*, *Pap. Bibl. Nat.* 197, 6, 2.

d. An order to make preparations for Pharaoh's arrival.

The three remaining lines of the fifth page of the *Koller*, after which the papyrus abruptly ends, contain the beginning of a long letter that is more completely preserved in *Anastasi IV* 13,8—17,9. A detailed consideration of this letter would here be out of place, and the translation of the three lines preserved in the *Koller* is given below only for the sake of completeness. The entire text will be dealt with later in connection with *Anastasi IV*.

5,5 The scribe Amenope writes ⟨to⟩ the scribe Paibēs. This letter is brought unto thee to say: — Take heed to have preparations made for Pharaoh, thy good lord, in fair and excellent order. Do not draw down punishment upon thyself. Look to it diligently, and beware! Do not be remiss! List of all the things that thou shalt cause to be provided. — Let materials be procured for the basket-makers consisting of reeds and and *is-r*-grass; and likewise . . . . (End).

Druck von August Pries in Leipzig.

Symbols, abbreviations etc..

Lacunae are always indicated by cross-hatching ▨; blank spaces in the original are specially noted; blank spaces in the publication have no significance.

Restorations are indicated both by square brackets [ ], and by cross-hatching ▨.

A vertical dotted line ⋮ indicates that the signs next following are not the Ms. continuation of what precedes, but are transposed from elsewhere.

Two such dotted lines ⁞⁞ indicate the omission, in the publication, of some signs or words contained in the original.

Sic above or below a sign implies that the Ms. reading is reproduced with special care, and that no doubt need be felt by the reader.

tr. " " " " " indicates traces suitable to the reading given.
pr. " " " " " " that the reading is probable, but not certain.

Words or signs in red in the original are underlined; verse-points • are always red unless otherwise stated.

In the textual notes the numbers used in quotation refer, wherever practicable, to page and line of the original manuscript: thus in this Part 2,5 would indicate Anastasi I, page 2, line 5; on the other hand p. 2, l. 5 would signify page 2, and line 5 of my publication of literary texts.

Hieratic signs in the notes have been traced from the original or from a good facsimile unless otherwise stated or clearly implied.

1. A satirical letter from one scribe to another,

transcribed from Pap. Brit. Mus. 10247 (Anastasi I) and from sundry other fragmentary manuscripts.

(1,1) Anastasi I — 1
?pr. a — (1) Ostr. Petrie — 2

sic — An. I — 3
(2) ?b — O. P. — 4

(1,2) sic — An. I — 5
O. P. — 6

sic c — An. I — 7
a. sic sic (3) — O. P. — 8

sic d — An. I — 9
e — O. P — 10

2.[a] [hieratic] — There is hardly room for ndnd before this. Read [hieroglyphs ?]

4.[b] [hieratic] = [hieroglyphs] ?

7.[c] [hieroglyph] corruptly for [hieroglyph], as (e.g.) in kd (Anast. I) 2, 1; stpw 2, 6.

9.[d] With dot, i.e. properly [hieroglyph].

10.[e] O. P. here interposes several sentences, see below p. 2 l. 2 et seqq.

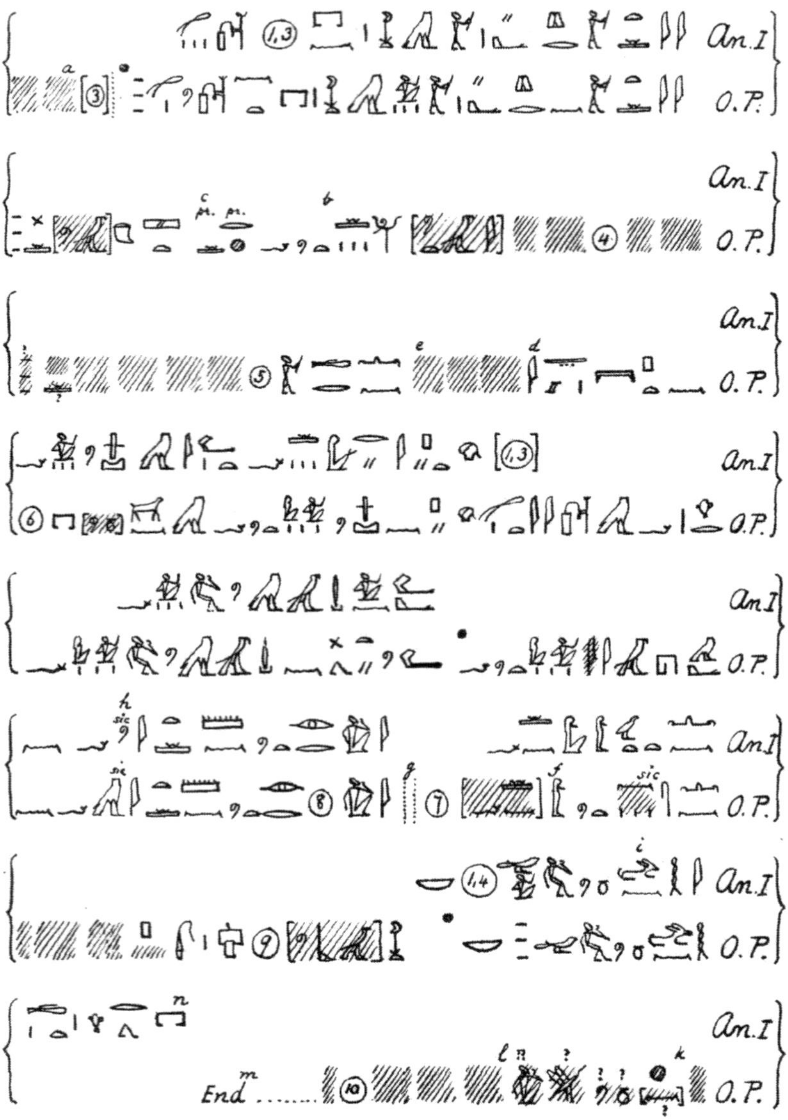

(1,3) An. I

a (3) O.P.

An. I

c pr. pr. b (4) O.P.

5

An. I

(5) e d O.P.

(1,3) An. I

(6) O.P.

An. I

O.P.

10

h sic An. I

sic (8) g (7) f sic O.P.

i (1,4) An. I

(9) O.P.

n An. I

15

End m .... (10) l ? ? ? ? k ? O.P.

2.[a] Immediately following upon m isy-f n sšw (above, p. 1, l. 10) O.P. has some sentences absent from An. I. Restore the first [sš3w m] i3wt-f ?

4.[b–c] Thus:- almost invisible, i.e. rḫ latter added over the suffix (tw-)f.

6.[d–e] hardly [hieroglyphs] or [hieroglyphs]

11.[h] Not m, though rather similarly made.

12.[f] Now scaled away, but seen by me. — [g] Here the words sb3y(t) etc., see above p. 1, l. 10 – p. 2, l. 2.

13.[i] [hieroglyph] i.e. properly [hieroglyph] (in), a common error.

15.[n] Possibly these words should not be separated from iḫwnw nb (above l. 13); but [see O.P.

16.[k] Doubtful traces. — [m] The recto may have had another line; the verso shows nearly illegible traces of two lines interpunctuated with red, which may belong to this text.

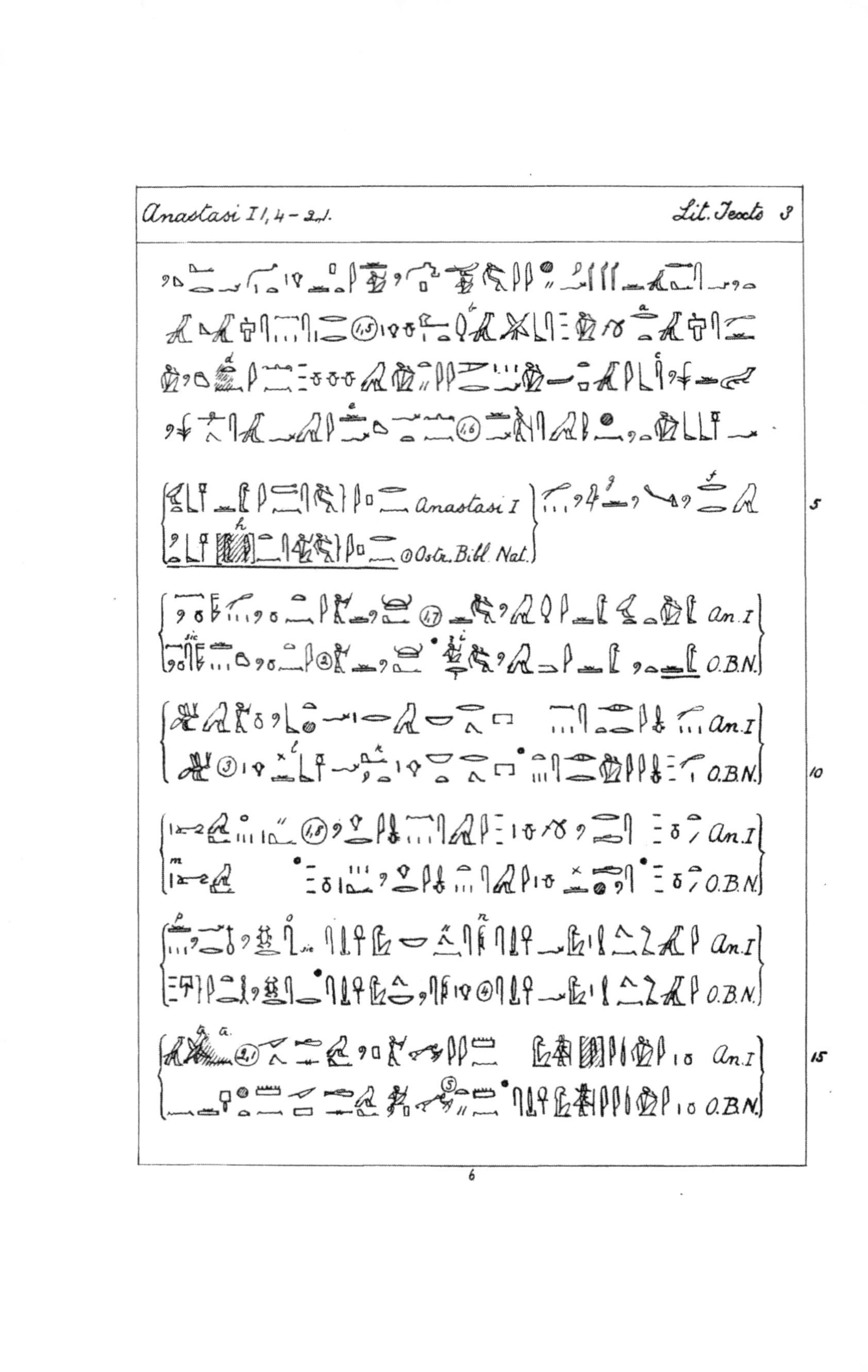

(1,5) (1,6)

Anastasi I
(1) Ostr. Bibl. Nat.

5

(1,7) An. I
sic (2) O.B.N.

An. I
(3) O.B.N.

10

(1,8) An. I
O.B.N.

sic An. I
(4) O.B.N.

(2,1) An. I
(5) O.B.N.

15

2.[a] Really [hieroglyph] under the influence of [hieroglyph]. [b] Read *sb(ꜣ)k*, cf. 41,3

3.[c] Corrupted from [hieroglyph], as often. [d] [hieroglyph] (indistinct trace); the [hieroglyph] like a [hieroglyph].

4.[e] Corrupted from *ft* [hieroglyph].

5[f–g] Or rather [hieroglyph]; clearly a corruption of [hieroglyph]; see too 12,1.

6[h] Sp(iegelberg)'s copy has indistinct traces.

8.[i] So Sp.; read [hieroglyph]?

10.[k] Read [hieroglyph] *sp·t*? [l] Read *ꜣbḫ*.

12[m] Sp's copy omits; so photogr.

13[n] Properly [hieroglyph] *šmꜥ*; contrast 3,4.5. [o] See facsimile. [p] Corrupt for [hieroglyph].

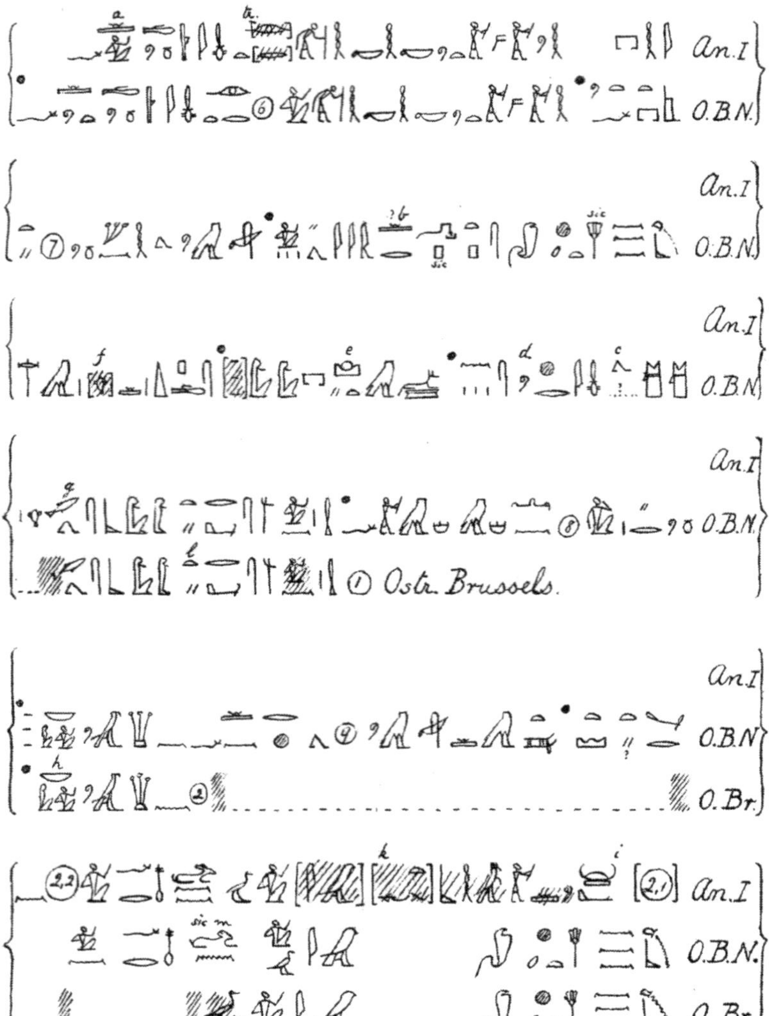
An. I
O.B.N.
An. I
O.B.N.
An. I
O.B.N.
An. I
O.B.N.
Ostr. Brussels.
An. I
O.B.N
O.Br.
An. I
O.B.N.
O. Br.

1.[a] [hieroglyphs] is for [hieroglyphs]; see above p. 1a, note 7[c].

4.[b] Sp. transcribes [hieroglyphs], but his hand-copy has a clear [hieroglyph] with an uncertain group following; the latter must contain the determinative [hieroglyph].

6.[c] Sp's hand-copy [hieroglyphs]. — [d] Sp. reads [hieroglyphs]; perhaps for [hieroglyphs]. — [e] Sp. [hieroglyphs] but the det. [hieroglyphs] are in favour of [hieroglyphs].

6.[f] Practically certain; the stroke of | [hieroglyph] clear in Phot.

8.[g] Clear in Phot.

9.[l] Hardly [hieroglyph].

12.[h] Hardly [hieroglyphs].

13.[i–k] The traces [hieroglyphic traces] make nhb certain; if the restoration nhb-[t] is right, the phrase wḥꜥ nhb-t belongs logically to mniti mds n pꜣ iḥ etc., above 1,8–2,1.

14.[m] So Phot., not [hieroglyph].

An. I
O.B.N.
O.Br.

An. I
O.B.N.
O.Br.

An. I
O.B.N.
O.Br.

An. I
O.B.N.
O.Br.

An. I
O.B.N.
O.Br.

2.[a] So too Phot.; doubtless an error for 𓏤, — [b] Sp.: "mit roter Tinte in 𓏤 korrigiert." cf. below p. 15 line 5.

5.[c] So apparently Sp.'s hand-copy.

11.[d] Not clear in the hand-copy; so Sp.'s transcription. — 𓏤 (so too Phot.) probably arose from a mistaken reading of the hieratic group for m<u>š</u>ᶜ, which has some points of similarity.

[a] An.I
sic O.B.N
(6) O.Br.

(2,5) [b] An.I
(14) O.B.N
End[c] O.Br.

An.I
(15) O.B.N

(2,6) An.I
sic [e] [d] O.B.N.

[f] An.I
O.B.N.

(2,7) An I
O.B.N.

An.I
sic sic (16) [h] O.B.N
End[i]]

1.[a] Properly [hieroglyphs], see above p. 3a, note 2[a].

4.[b] Without dot, like [hieroglyph].

6.[c] It is uncertain whether there were more lines, now lost.

10.[d] So Sp.'s hand-copy and transcription. — [e] [hieroglyph] a corruption out of a hieratic [hieroglyph].

11.[f] [hieroglyph] is for [hieroglyph], see above p. 1a note 7[c]; for stpw see below 5,1 end; 7,8 end.

13.[g] [hieroglyph] is corrupted out of hieratic [hieroglyph]; read isf.t

16.[h] Sp.'s hand-copy [hieroglyphs] — [i] Finally the colophon: [hieroglyphs]
"on the day when the scribe Nfr-ḥtp came."

t. An. I

(2,8)

(3,1)

pr. a ?a.

pr. b (3,2) pr. 5

e d c sic

a. g (3,3) f sic ?

k i sic h

m sic l (3,4)

pr. sic sic Anastasi I 10

a. a. (2) sic Ostr. Queen's Coll. Oxford

An. I

? n (3) sic O.Q.C.

An. I

q p (4) o O.Q.C 15

An. I

(5) r O.Q.C

4.[a] At beginning, [hieroglyphs] would be a possible reading. The space is rather small for [m3]3, and the final 3 by no means certain.

5.[b] Confusedly mounted and not quite certain.

6.[c] Scanty, but sufficient, traces. — [d-e] [hieroglyphs] ; not hrp(w).

7.[f] Perhaps no lacuna. — [g] Cf. above 2,2.

8.[h] Cf. 8,6 below. — [i] The correction [hieroglyphs] at top of page belongs here, if not to wpwt in 3,6 ad init. [k] Cf. below 4,4.

9.[l] The trace of m very curiously made, possibly a later addition. — [m] Quite clear in the original.

13.[n] Probably merely reconstructed from [hieroglyphs] wnm.

15.[o] Error for [hieroglyphs]. — [p] Original [hieroglyphs], 3 gs and point added in red above. — [q] False analogy with iʿ "to wash".

17.[r] Or II ?

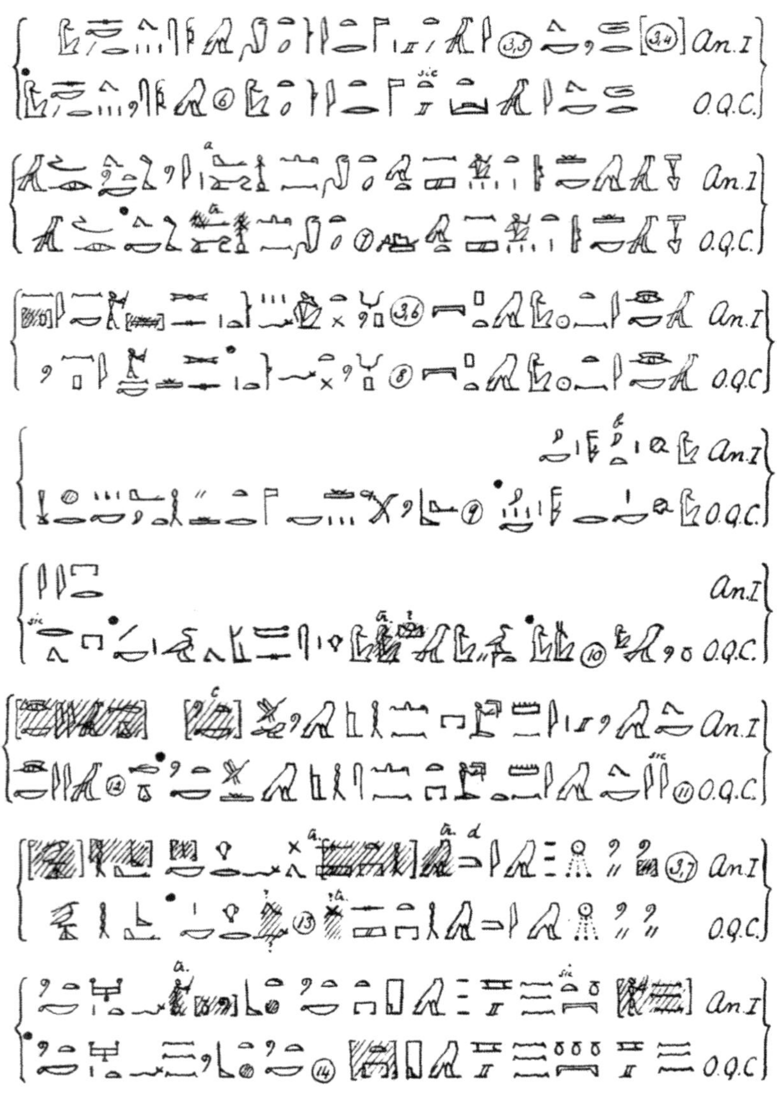

(3,4) (3,5) An. I
(6) O.Q.C.

a An. I
a (7) O.Q.C.

(3,6) An. I — 5
(8) O.Q.C.

b An. I
(9) O.Q.C.

An. I
sic a (10) O.Q.C.

c An. I — 10
(12) sic (11) O.Q.C.

a a d (3,7) An. I
a (13) O.Q.C.

a sic An. I
(14) O.Q.C. — 15

3.[a] As often, [hieroglyphs] here replaces [hieroglyphs].

7.[b] [hieroglyph], erroneously for [hieroglyph].

10.[c] Cf. 4, 5.

12.[d] [hieroglyph]

(3,8) An. I

(16) (15) O.Q.C.

An I

(19) (18) (17) O.Q.C

(4,1) An. I 5

(21) (20) O.Q.C.

An. I

End (22) O.Q.C

(4,2)

10

(4,3)

(4,4)

15

(4,5)

2.[a] The first ꜥ in this spelling of iwꜥḥ is derived from iwꜥ "inherit"; the second from iꜥ "wash".

6.[b] Very faint, but probable.

7.[c] A tiny trace, suitable to [hieroglyph]. — [e] Cf. Ṯꜣ at end of line, but the name is elsewhere spelt with [hieroglyphs].

8.[d] The ostracon, which is complete, ends with these red signs.

10.[e-f] [hieroglyphic traces] — Sethe (1901) suggested snk·k irt·t, but ḏ in mnḏ is certain, and mn probable, though the det. is quite obscure. A possible trace of m before mnḏ — 10.[j] After [hieroglyph] a small misplaced fragment.

11.[g] Or 11. — [h] Hardly [hieroglyphs], since [hieroglyphs] is written 1,1; 4,5. — 11.[i] [hieroglyphic traces]

12.[k] A corruption of wšbti; [hieroglyph] is certain.

12.[l] – 13.[m] Corrupted from [hieroglyphs]

14.[n] For [hieroglyph] cf. 9,9.

16.[o] Not wbꜣ, which would require [hieroglyph]; not space enough for sbꜣ. Nor can we have here a writing of dbꜣ-t; [hieroglyphs] is not found as a writing of [hieroglyphs]. There is a trace high up at beginning, but it is dull and may be a smudge.

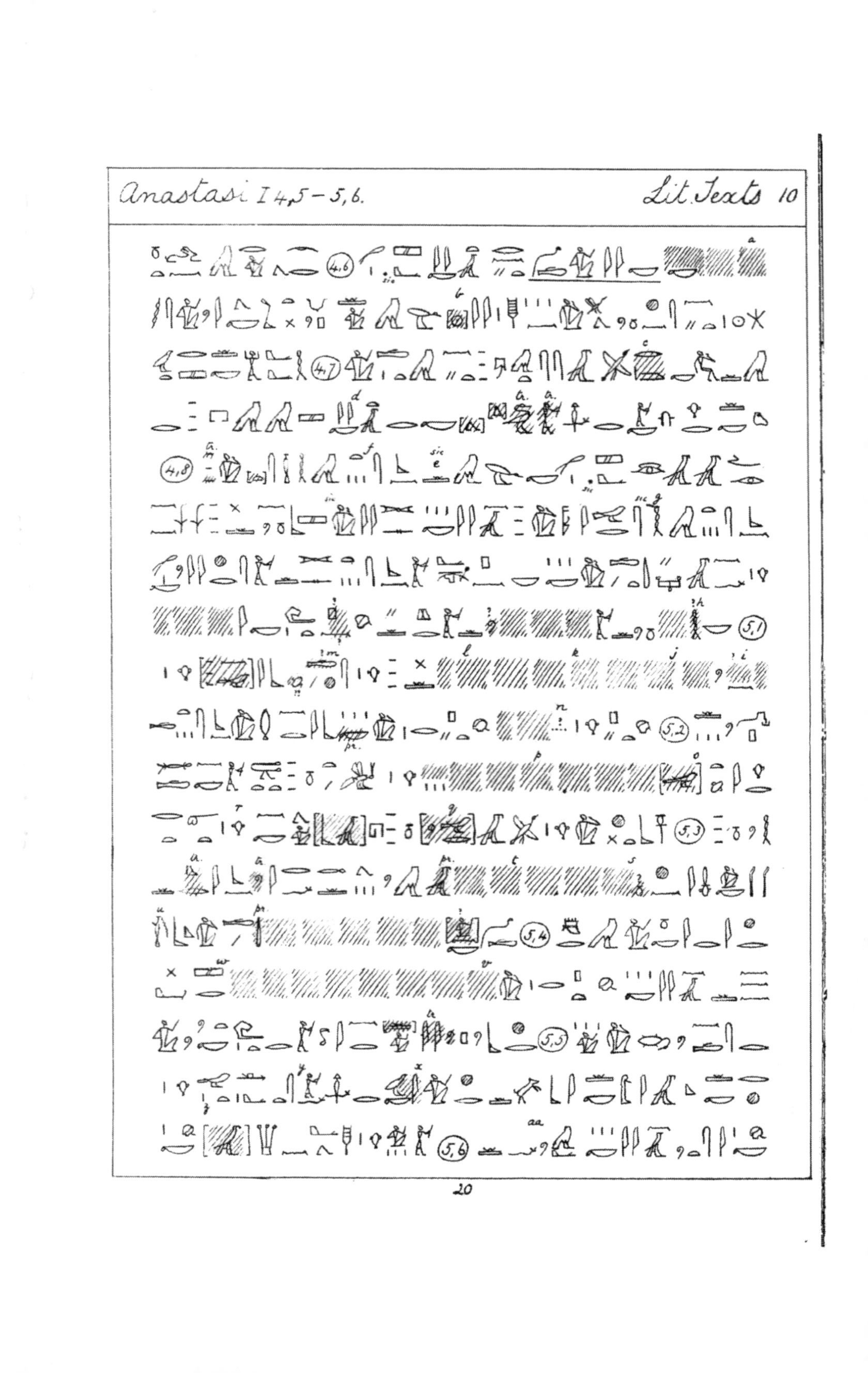

(4,6) (4,7) (4,8) (5,1) (5,2) (5,3) (5,4) (5,5) (5,6)

1.[a] [hieroglyph]

2.[b] Cf. 17,1 end.

3.[c] Damaged and displaced, but certain.

4.[d] Corrupted from [hieroglyphs].

5.[e] Read gm-i. – [f] Corrected out of [hieroglyphs].

6.[g] [hieroglyph] – The stroke may be meant to imply that ḥ and š should be transposed; read šḥwri.

8.[h] A vertical sign.

9.[i] See facs.; utterly confused, with superimposed fragments.

[j-k] Here and below 5,2.3 a large misplaced fragment, fitting into 5,5.6.7; see facs.

[l] Here the misplaced beginning of a page, recognized by Sethe as belonging to 9,1.

9.[m] Facs. accurate; uncertain.

10.[n] [hieroglyph] – not [hieroglyph].

11.[o-p] See the note 9[j-k] above.

12.[q] Cf. Anast. IV 7,4; 16,1. – [r] Under [hieroglyph] a misplaced fragment.

13.[s-t] See the note 9[j-k] above.

14.[u] Properly only [hieroglyph]; differently below (e.g.) 7,2.

15.[v] Undecipherable fragments. – [w] Perhaps a word [hieroglyphs] cf. 5,8.

17.[x-y] See the note 9[j-k] above, and the facs. on p. 11a. – [z] Corruptly for [hieroglyphs].

18.[aa] Emend [hieroglyphs], cf. below 9,1 and variants, all corrupt.

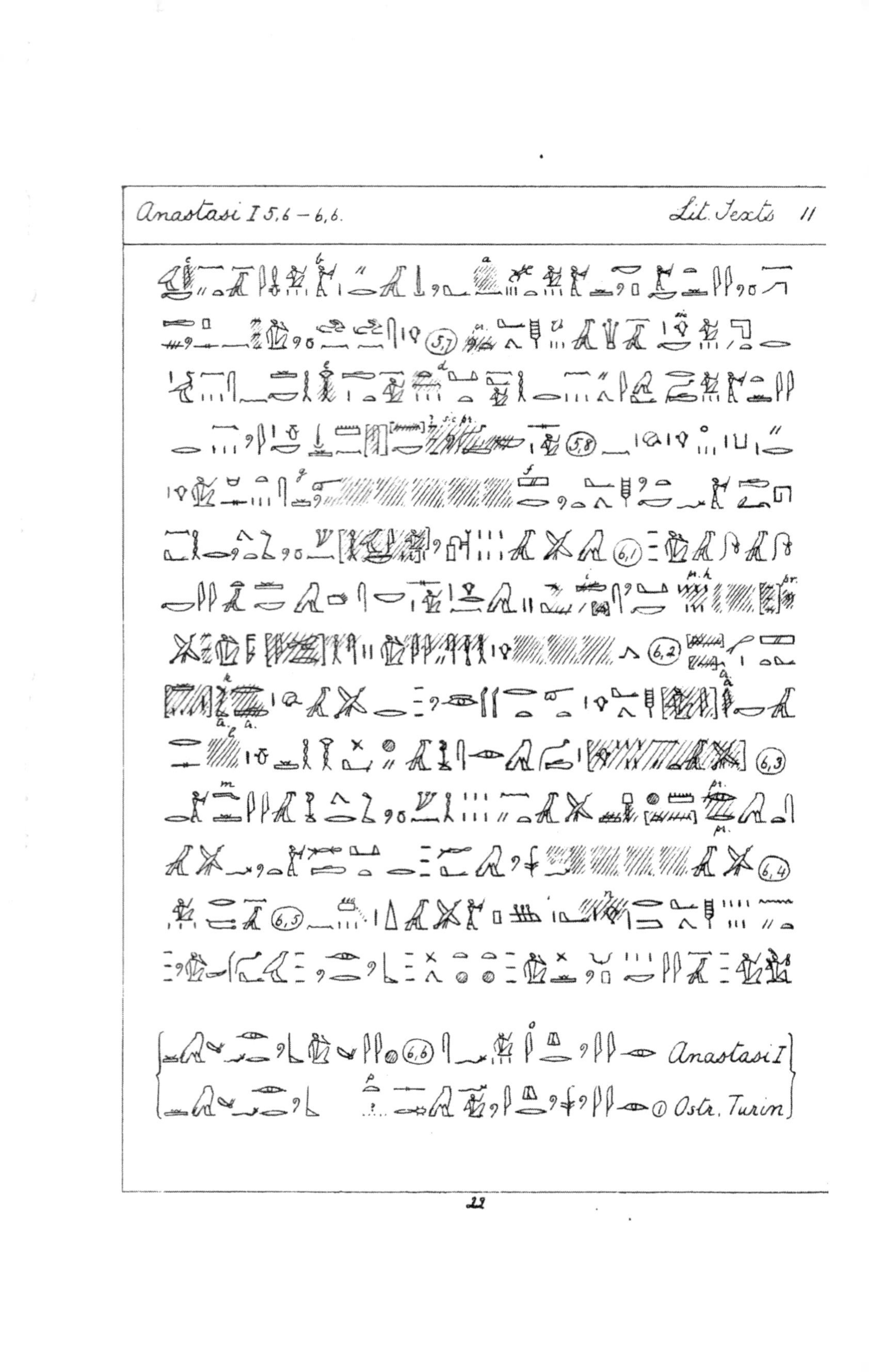

c b a

(5,7) a. b m

e d

? sic pr. (5,8)

g f

(6,1)

i p.h pr.

(6,2)

k a. a.

a. l a.

(6,3)

m pr.

pr.

(6,4)

(6,5) n

o

(6,6) Anastasi I

p

① Ostr. Turin

1.[a] [hieroglyph] ; [hieroglyph] does not suit the context.

1.[b–c] See above p. 10a, note 9[j–k]; the accompanying facsimile represents the misplaced fragment restored to its proper position. On the right the papyrus has been unequally stretched.

3.[d–e] See the note 1[b–c]

5.[f] See p. 10a, note 15[w]. –[g] Perhaps corrupted from m bh [hieroglyph].

5
6
7

7.[h] Cf. 6,4. –[i] Traces suit well; cf. 7,8.

9.[k] Cf. 28,8; so already Goodwin, in a Ms. note in his copy of the Select Papyri, now in my possession.

10.[l] The traces do not suit [hieroglyph], and there is hardly room for [hieroglyph].

11.[m] [hieroglyph]

13.[n] The upper portion of these signs has become detached, and is mounted too far towards the left.

15.[o] The proper name [hieroglyph].

16.[p] Not [hieroglyph], see the facsimile of Rossi.

An. I
O.T.

An. I
O.T.

An. I
O.T.

An. I
O.T.

An. I
O.T.

An. I
O.T.

An. I
O.T.

An. I
O.T.

1.[a] Read , cf. for above p. 3a, note 4[c].

5.[b]  an unusual form, cf. on the ostracon Berlin 12337.

6.[c] ; thus in my notebook, an unintelligible corruption.— [d] My notebook gives

9.[d] Slightly displaced in original.

14.[e] An illegible sign above ; before this Rossi gives , my notebook .

15.[f] So facsimile; the signs now destroyed in original.

16.[g] Not quite certain, see Rossi's facsimile.

(7,2) An. I
O.T.

An. I
O.T.

An. I
O.T.

(7,3) An. I
(9) O.T.

An. I
End. O.T.

(7,4) An. I

(7,5)

(7,6)

(7,7)

2.[a] Probably so, not [hieroglyph].

3.[b] Emend [hieroglyph].

4.[c] See Rossi's facsimile; a corruption of bsy.

6.[d] Probably for ptri-i.

7.[f] [hieroglyph] is superfluous.

8.[e] Certain, but confused by meaningless smudges.

10.[h] An uncertain stroke, see facsimile. —[i] Unintelligible signs, see facsimile.

11.[k] A trace (see facsimile) which might belong to [hieroglyph]. —[l] Read [hieroglyph] ?

14.[m] [hieroglyph]; the upper parts of the signs, on a separate fragment, are however very possibly misplaced.

16.[n] The signs Ddw are superfluous and may be an attempted etymology of Dhwti, Thoth.

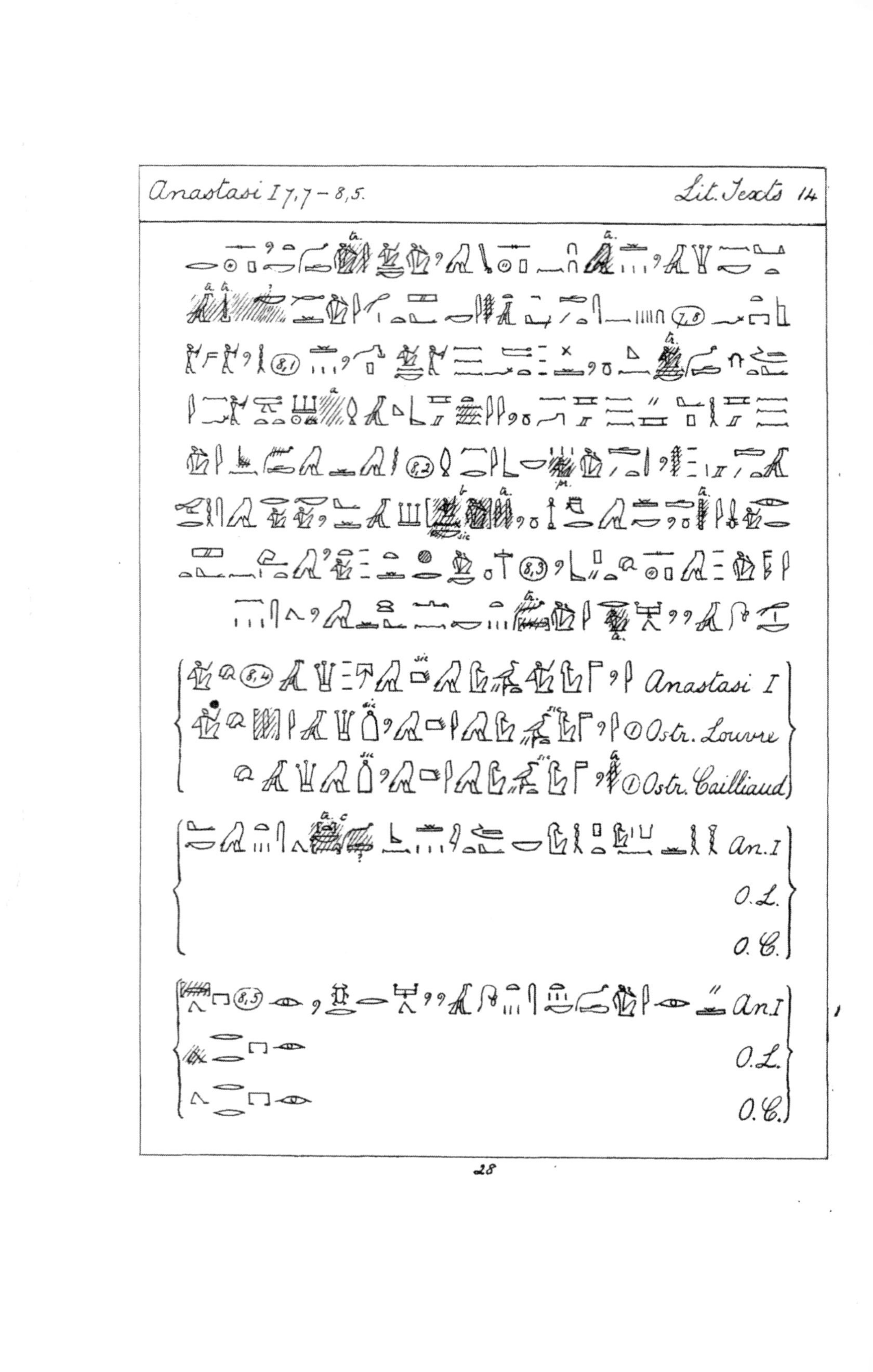

(7,8)

(8,1)

(8,2)

(8,3)

(8,4) Anastasi I

① Ostr. Louvre

① Ostr. Cailliaud

An. I

O. L.

O. C.

(8,5) An. I

O. L.

O. C.

4.[a] Possibly restore [hieroglyph].

6.[b] Cf. above 5,3–4; the first person is made probable by the termination -ti.

12.[c] See facsimile: [hieroglyph] is very probable, [hieroglyph] quite uncertain; the left end of a sign seems to suit [hieroglyph], and a point may belong to [hieroglyph] below it.

An. I
O.L.
O.C.

An. I
O.L.
O.C.

An. I
O.L.
O.C.

An. I
O.L.
O.C.

An. I
O.L.
O.C.

1.[b] There is sufficient room for the m.

3.[a] So apparently the facsimile.

4.[f] A word like [hieroglyphs] is perhaps lost after nb.

6.[c-d] Later added above the line. — [e] M3ˁ-t m ỉ3 omitted because of the homoio=[=archon]

7.[g] Quite doubtful.

11.[h] The omission of [hieroglyph] in both O.L. and O.C. proves their close relationship.

13.[i] Emend [hieroglyphs].

(8,7) An. I
(4) O.L.
sic (4) O.C.

b
An. I
O.L.
sic O.C.

(8,8) An. I
(5) O.L.
(5) O.C.

c
An. I
d
O.L.
sic
e O.C.

i h g (9,1) An. I
f (6) O.L.
(6) O.C.

5
10
15

1.[a] M is superfluous.

4.[b] [hieroglyph] corrupted out of [hieroglyph], cf. above p. 3a, note 3c.

10.[c] Probably nothing lost.

11.[d]–14.[f] Bw rh-f erroneously repeated.

12.[e] K for f by assimilation with the suffix of dd-k.

13.[g] See above p. 10a, note 9l.—[h] A low sign.—[i] [sign] later added in red.

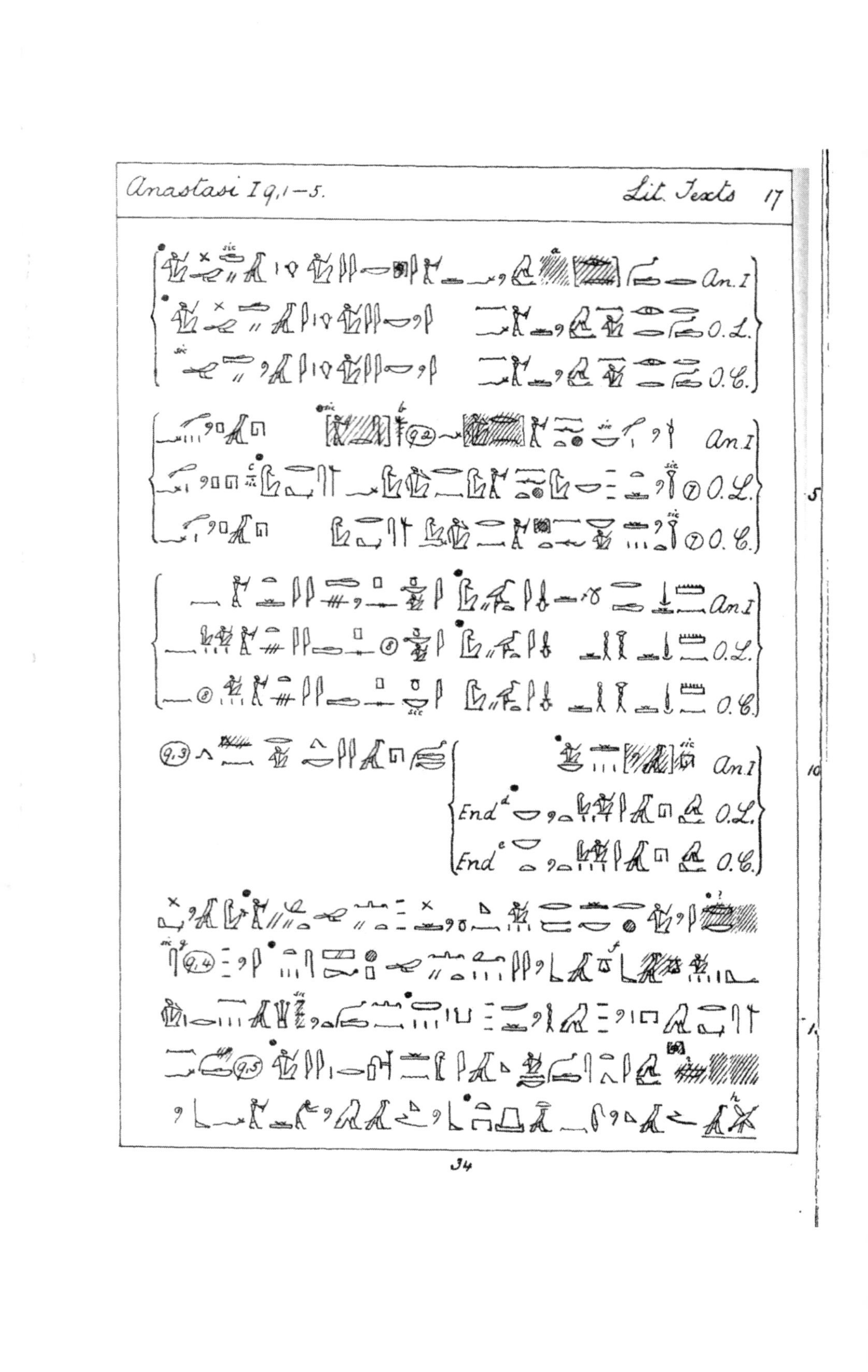

sic a An. I

O.L.

sic O.C.

sic b (9,2) sic An. I

c sic sic (7) O.L. 5

sic (7) O.C.

An. I

(8) O.L.

(8) sic O.C.

(9,3) sic An. I 10

End d O.L.

End e O.C.

?

sic g (9,4) f

sic

(9,5)

h

1.[a] At top a trace (see facsimile), perhaps [hieroglyphs] — For the word mʿwnf cf. above [p. 16a, note 18[aa]

4.[b] Damaged but still recognizable.

5.[c] A meaningless dash.

11.[d] At end a date [hieroglyphs] sic "third month of winter, day 29."

12.[e] Colophon: [hieroglyphs] "By the fan-bearer on the right of the king, the Governor of the Town and Vizier To."

14.[f] Read gꜣb [hieroglyph]. — [g] No sign lost before š, which is perhaps superfluous.

17.[h] In red above the line.

(9,6) (9,7) sic a (9,8) sic sic sic (9,9) b sic c

Anastasi I

Ostr. Berlin

(10,1) d An. I

O.B.

e An. I

O.B.

f An. I

O.B

h (10,2) An. I

O.B

4.[a] Ḥpt [hieroglyph] erroneously determined by the entire group [hieroglyphs] ḥrw.

7.[b] A meaningless dash (Füllstrich).

8.[c] This restoration fills, or very nearly fills, the lacuna; cf. Anast. IV 9,7-8.

10.[d] Original [hieroglyphs]; w first added in red above line, then corrected into the text in black.

12.[e] For isw cf. above 9,8.

14.[f] After sw [hieroglyph] has been omitted.

16.[g] Not room for any restoration containing [hieroglyphs].

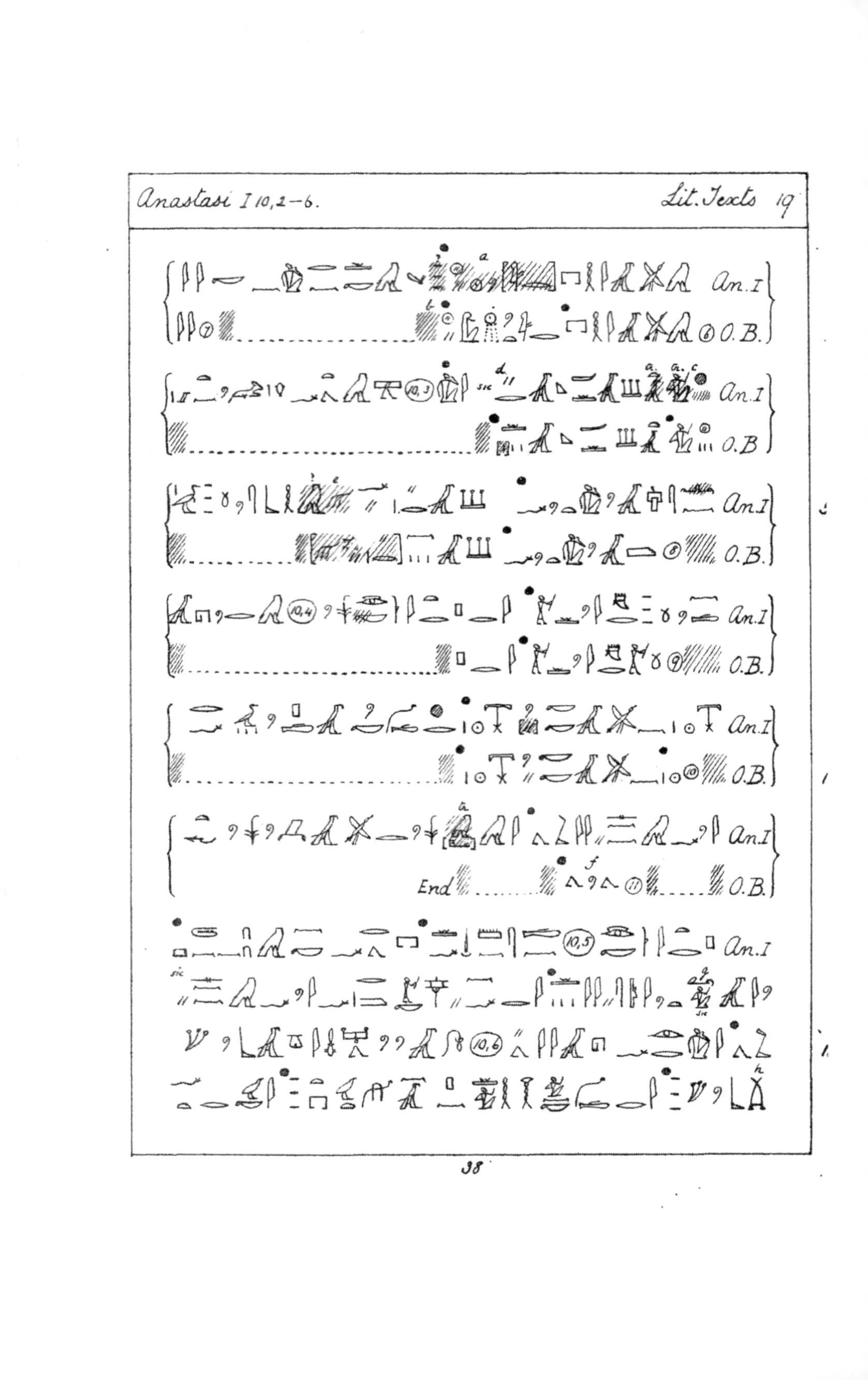

An. I
O.B.

sic An. I
O.B.

An. I
O.B.

An. I
O.B.

An. I
O.B.

An. I
End O.B.

An. I

sic

sic

1.[a]  ; the sign after [hieroglyph] is doubtful.

2.[b] [hieroglyphs]

3.[c] Under [hieroglyph] a small sign, perhaps [hieroglyph] or [hieroglyph] ; [hieroglyph] and [hieroglyph] are practically certain.

3.[d] [hieroglyph] with a vacant space following; this can hardly be read otherwise than <[hieroglyph]>[hieroglyph], in spite of O.B.

5.[e] [hieroglyphs] : [hieroglyph] has a quite abnormal form, but for the word cf. 24,1–2 ; before hbs a tail belonging to m or w.

12.[f] It is not clear to what word in Anastasi this iw corresponds.

14.[g] Read [hieroglyph], as in 17,5.

16.[h] [hieroglyphs]

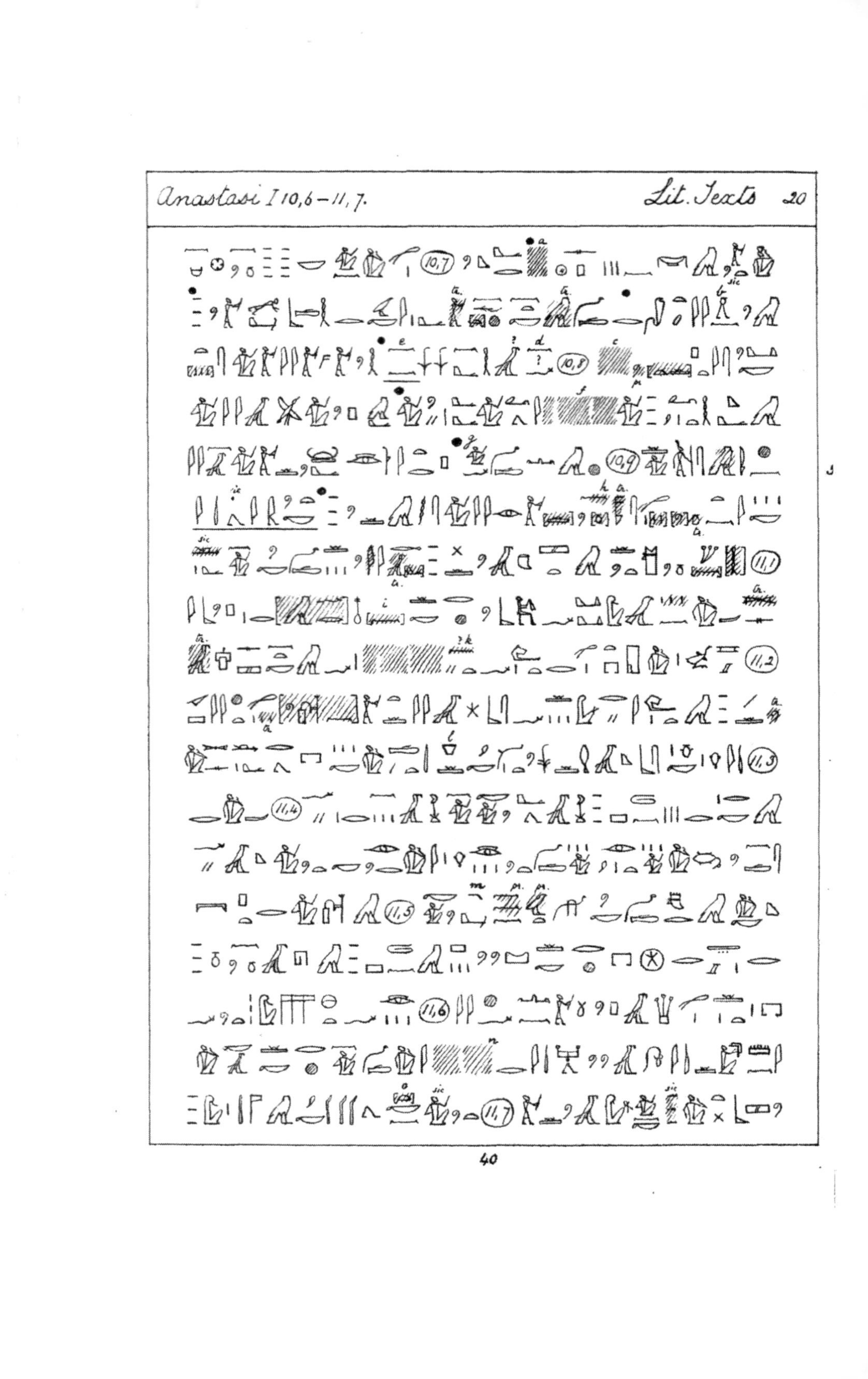

(10,7) (10,8) (10,9) (11,1) (11,2) (11,3) (11,4) (11,5) (11,6) (11,7)

1.[a] Something is lost after sp; the tiny trace in facs. is on a misplaced fragment.

2.[b] Read [hieroglyphs].

3.[c] At least something probably lost. —[d] [hieratic signs] meaningless signs, as they stand; probably a corruption of nfi.

4.[f] A verse-point might be lost after kꜥḥ-i. 3.[e] [sign] added in red above the line.

5.[g] A corruption of [hieroglyphs].

6.[h] [hieratic signs]; [sign] is nearly certain; ksn or sksn should be read, this being often contrasted with nḏm.

8.[i] A small low lacuna, which undoubtedly contained a sign; perhaps read [hieroglyphs].

9.[k] See facsimile; [sign] is far from certain.

11.[l] [hieratic sign]; [sign] as usual like [sign]; instead of [sign] the Ms. gives [sign].

14.[m] For the determinative cf. 23,3.

17.[n] Illegible traces, see facsimile.

18.[o] k and n are so close to one another that a sign must have stood above them.

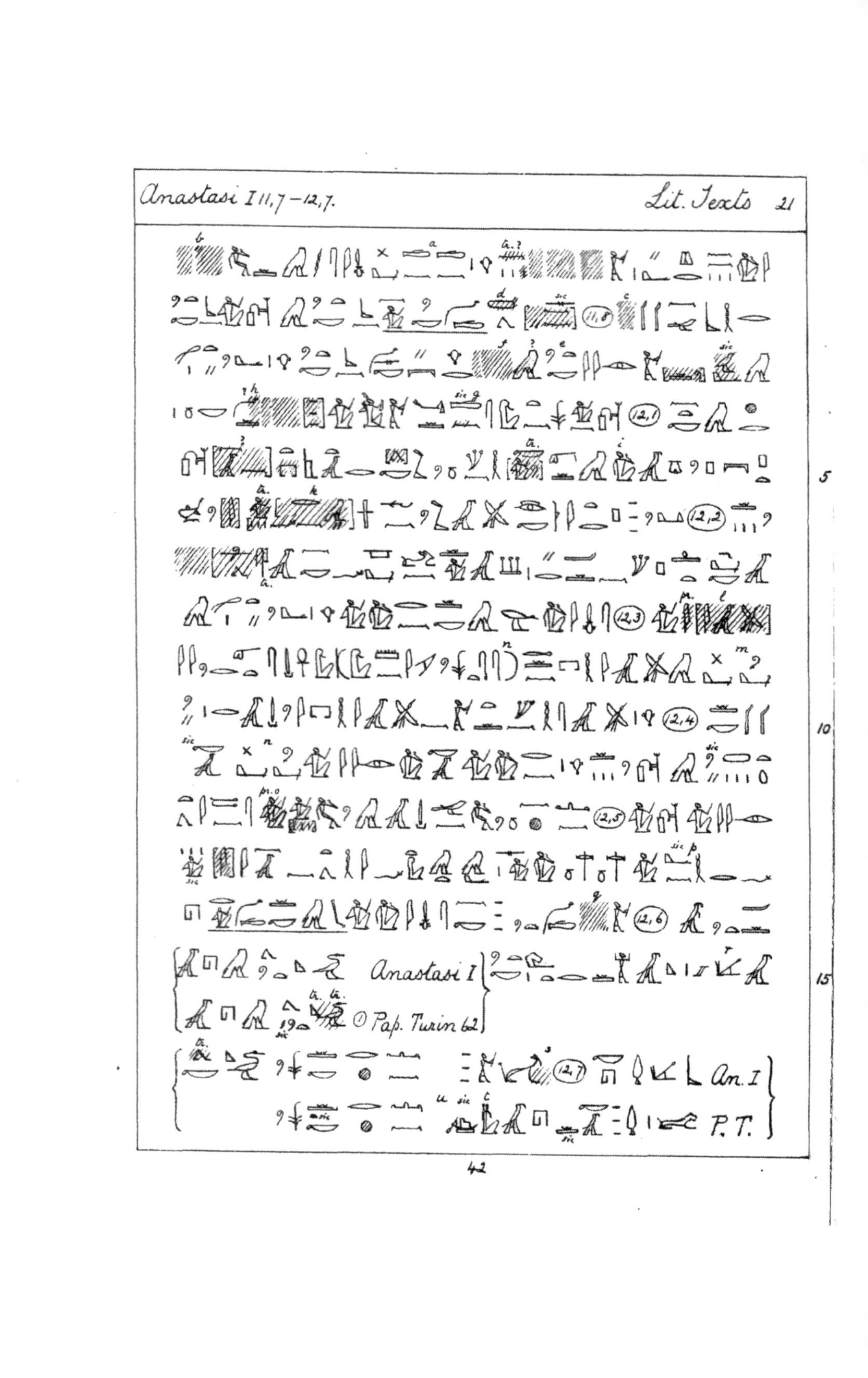

(11,8) (12,1)

5

(12,2) (12,3) (12,4)

10

(12,5) (12,6)

Anastasi I

15

Pap. Turin 62

(12,7) An. I

P. T.

1.[a] Sic, but tntn or inrn might be understood. — [b] [hieroglyphic traces] ; the last sign may be [sign].

2.[c] Very little can be lost. — [d] Cf. (e.g.) 16,7.

3.[e] Read iry-k tw m (?) cf. page 11 line 16 above. — [f] [hieroglyphic traces] ; the part above the break is probably a superimposed fragment ; the reading m [sign] hry is quite impossible.

4.[g] For [hieroglyphs], cf. p. 3a, note 5 f–g.

5.[i] Corrupted from [hieroglyph] ?   4.[h] [hieroglyphic traces]

6.[k] Suggested by Erman; there is a trace of [sign] and also of tail of [sign].

8.[l] Confused traces.

9.[m] Cf. the title [hieroglyphs] Mes N17 ; in hieratic, Pap. Leiden 350 verso, 3,34 ; the reversal of the groups here perhaps erroneous.   [n] Original [hieroglyphs] with [sign] later erased.

11.[n] Cf. note 9[m].

12.[o] Much more probable than [sign].

13.[p] Read [hieroglyphs], cf. 28,2 ; the scribe was thinking of [hieroglyphs]. The determinatives of stni ought to be [hieroglyphs].

14.[q] Puzzling traces, see facsimile ; not merely [hieroglyphs].

15.[r] It seems likely that [hieroglyphs] (cf. Pap. Turin 1), as lectio difficilis, was the original version, [hieroglyphs] being an easy corruption of [hieroglyphs] due to the influence of hꜣb "plough" and hꜣ(b)bni "ebony"; the absence of a variant with "ıı" makes it improbable that hꜣbni was meant.

17.[s] Probably no sign is lost at beginning of line.

18.[t] [sign] rather like [sign], whence [sign]. — [u] Correction above line: [hieroglyphs].

An. I
P.T.

(12,8) An. I
P.T.

sic a. a. An. I
b sic a sic P.T.

5

(13,1) An. I
P.T.

An. I
P.T.

10

d pr. c (13,2) An. I
P.T.

pr. pr. e An. I
P.T.

f (13,3) An. I
pr. a. P.T.

15

6.[a] See facsimile; the first [hieroglyph] very badly made. — [b] [hieroglyph] is here omitted.

11.[c] Dittograph due to change of line. — [d] Read perhaps [hieroglyphs] or the like.

13.[e] Emend idd [hieroglyph].

15.[f] If, as is probable, the reading nḥb-k tw "thou yokest thyself" is correct, the det. [hieroglyph] will be borrowed from nḥ-t, the first step towards the reading of P. T.

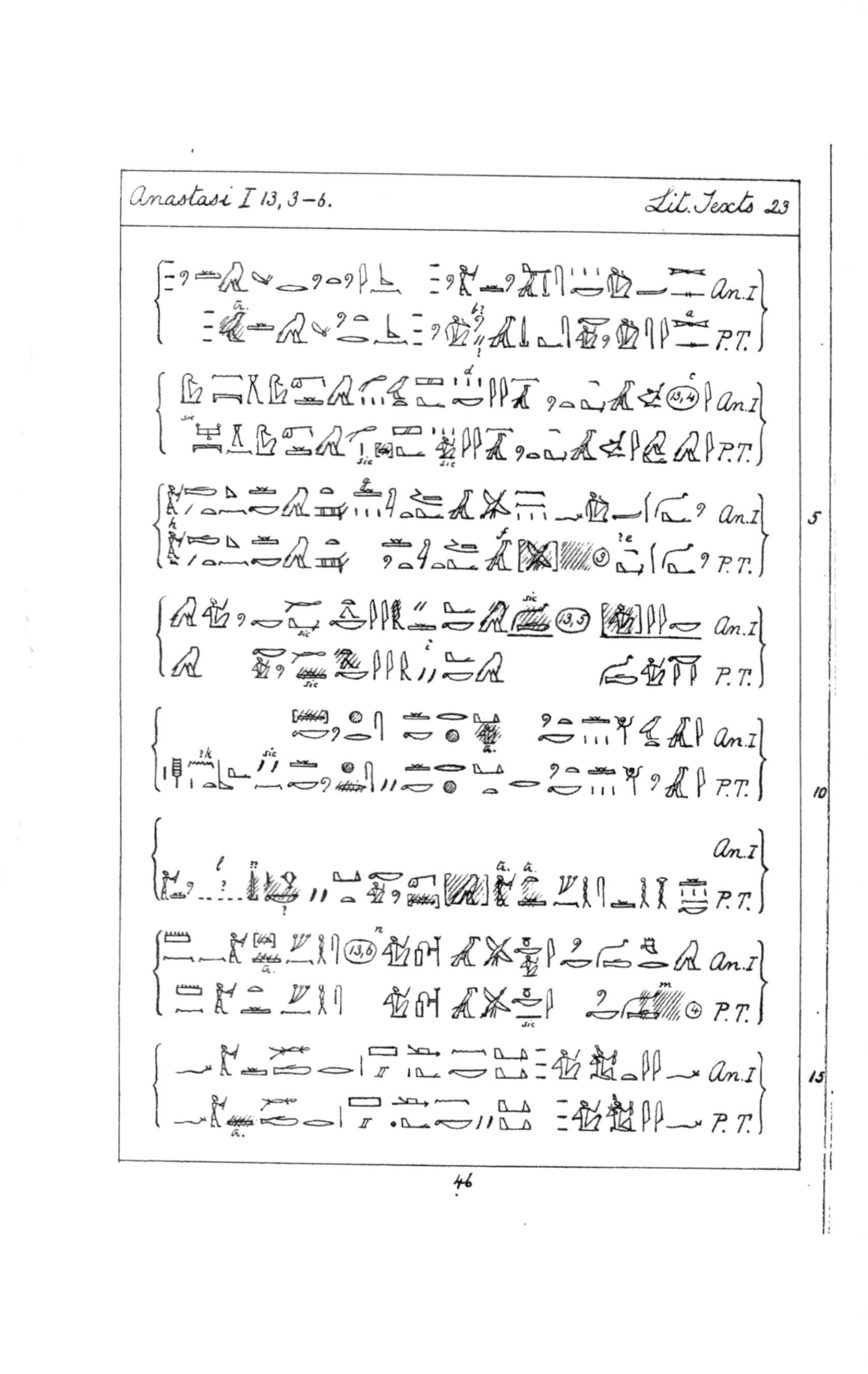

An. I
P. T.

(13,4) An. I
P. T.

An. I
(3) P. T. 5

(13,5) An. I
P. T.

An. I
P. T. 10

An. I
P. T.

(13,6) An. I
(4) P. T.

An. I
P. T. 15

2.[a] Very badly made, see facsimile. — [b] See facsimile.

3.[c] There is a hole after i, but its shape and size suggest that nothing is lost. — [d] Probably read nsi-i as P.T.; cf. above 4, 7.

5.[g] Read msˁt(w) with P.T.

6.[e] See facs., perhaps in part belonging to the earlier erased text. — [f] The ꜣ is small, and may belong to a later added pꜣ; there is hardly room in the lacuna for [ ]. — [h] See facs.

8.[i] Probably here for ?, as in ꜥḫ-k(wi) below l. 10.

10.[k] ; approximately thus, i.e. corrected from .

12.[l] Confused and injured signs.

13.[n] The line is shorter than the others, but probably nothing is lost at the end.

14.[m] It is very doubtful whether m dr stood at the beginning of the line.

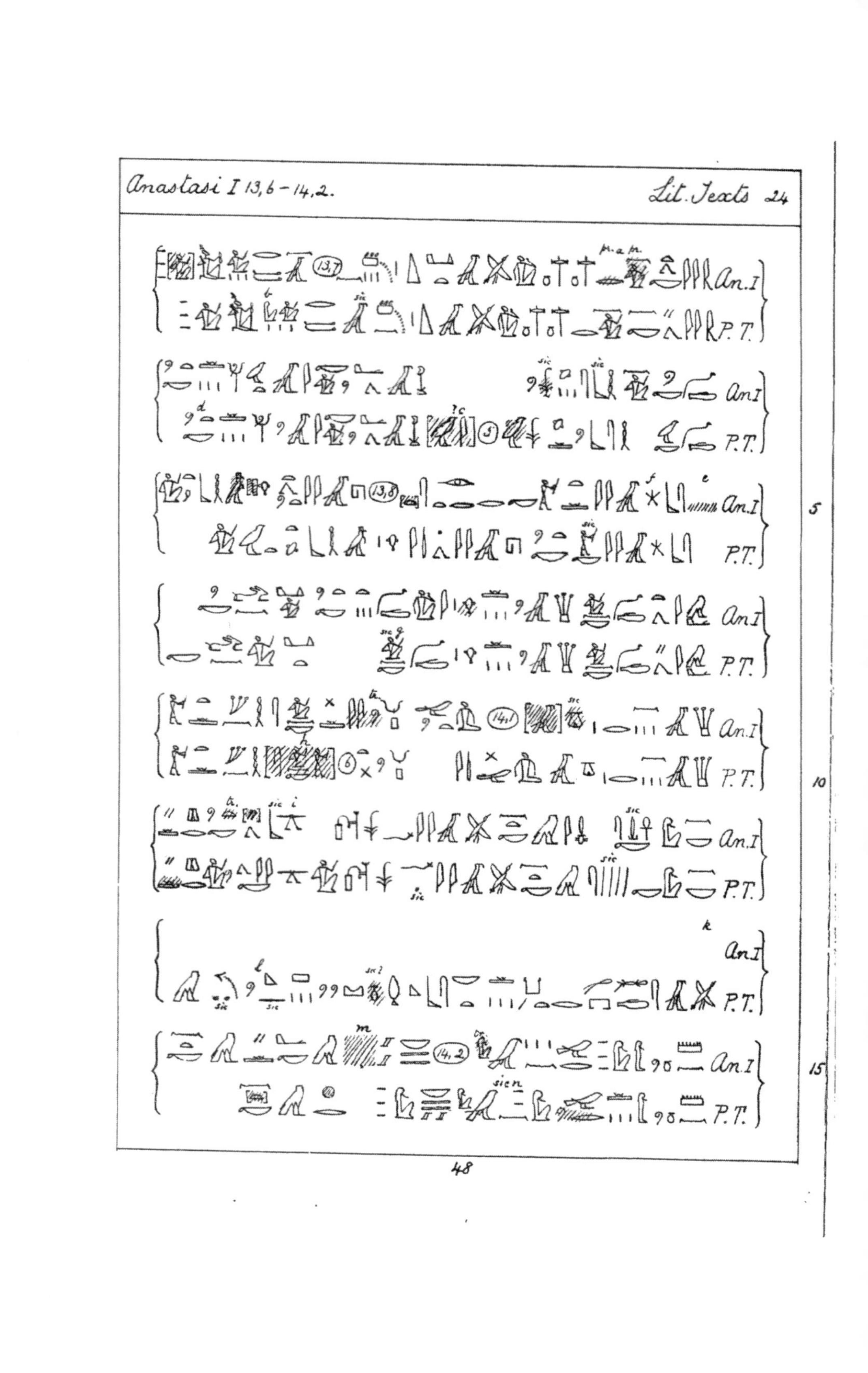

An. I
P.T.

An. I
P.T.

An. I 5
P.T.

An. I
P.T.

An. I
P.T. 10

An. I
P.T.

An. I
P.T.

An. I 15
P.T.

1.[a] Very probable; the original is here in disorder.

2.[b] Without a dot, i.e. like [hieroglyph].

4.[c] The other lines show that a word must have been lost here. If not tw, emend 3s "quickly", cf. 14,6.

[d] The [hieroglyph] is written over ‖

5.[e] A low lacuna, in which it appears something must have stood. —

[f] [hieroglyph] stands above a deleted sign.

8.[g] For [hieroglyph].

10.[h] The restoration may be a little too big for the lacuna; perhaps omit [hieroglyph].

11.[i] The upper part of these signs has become detached, and been pushed too far to right.

13.[k] An. I reads mnw wrw immediately after hr, which makes some sense; but it is probable that the archetype had the words here given by P. T.

14.[l] See facsimile; emend [hieroglyph]: cf. p. 26a, note 4[b].

15.[m] At top on left of lacuna, a trace like the corner of [hieroglyph] or [hieroglyph].

16.[n] The lowest of the plural strokes is lengthened to represent [hieroglyph].

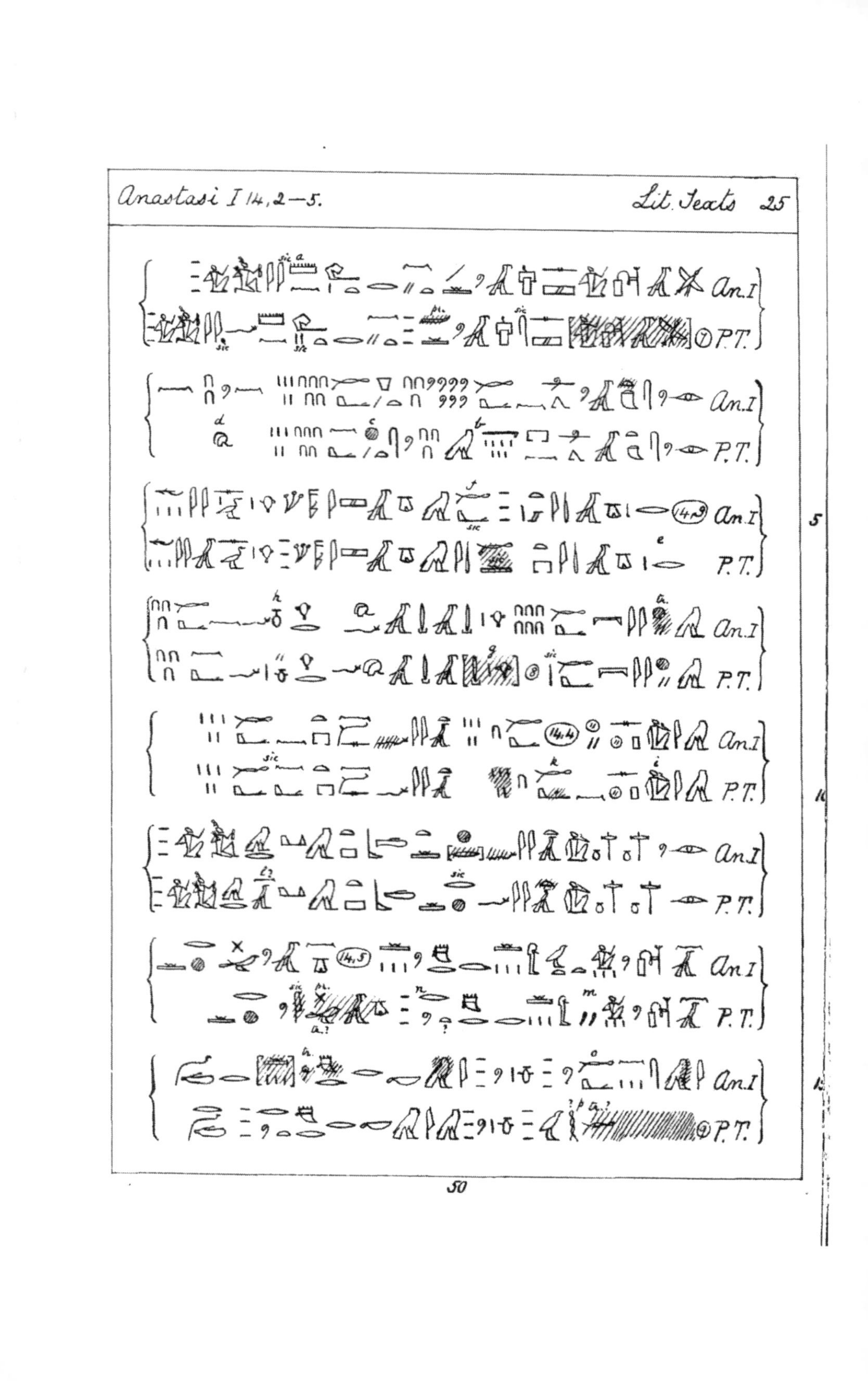

An. I
P.T.

An. I
P.T.

An. I 5
P.T.

An. I
P.T.

An. I
P.T. 10

An. I
P.T.

An. I
P.T.

An. I
P.T. 15

1.[a] Emend mnfy-t; the same corruption below 17,3.

4.[b] [hieroglyph] was probably intended by the scribe for [hieroglyph] sp-t 7, but is doubtless a corruption of [hieroglyph]; the following [hieratic] = [hieroglyph] is obviously derived from [hieratic] 700; see An. I.

4.[c] The hieratic sign here, (see above p. 20a, note 11[l]) does double duty for [hieroglyph] and for [hieroglyph].

4.[d] [hieroglyph] here is an obvious and easy corruption of [hieroglyph]. The facsimile gives a slanting stroke at bottom of the preceding [hieroglyph], which may be the last trace of [hieratic] 100. The reading of An. I is [clearly correct.

5.[f] Read [hieroglyph].

6.[g] Small like [hieroglyph].

7.[h] [hieratic], corrupt for [hieroglyph] cf. P.T.

8.[i] There may have been a numeral (60 ?) before hr, as the lacunae at the beginnings of lines grow bigger towards bottom of page, see esp. P.T. 10.11. However the stroke in the pre-ceding [hieroglyph] is quite unusual, if here a determinative.

10.[i] The abbreviated form, which is rare in literary texts. See facsimile.— [k] [hieroglyph] over a deleted [hieroglyph]?

12.[l] [hieratic], for [hieroglyph] or [hieroglyph], but like n3.

14.[m] For 9 [hieroglyph]. — [n] See facsimile, and cf. next line of P.T.

15.[o] Read [hieroglyph], cf. above note 5[f].

16.[p] My notebook gives [hieratic]; possibly there was room for Im-sn before mh, see above note 8[i].

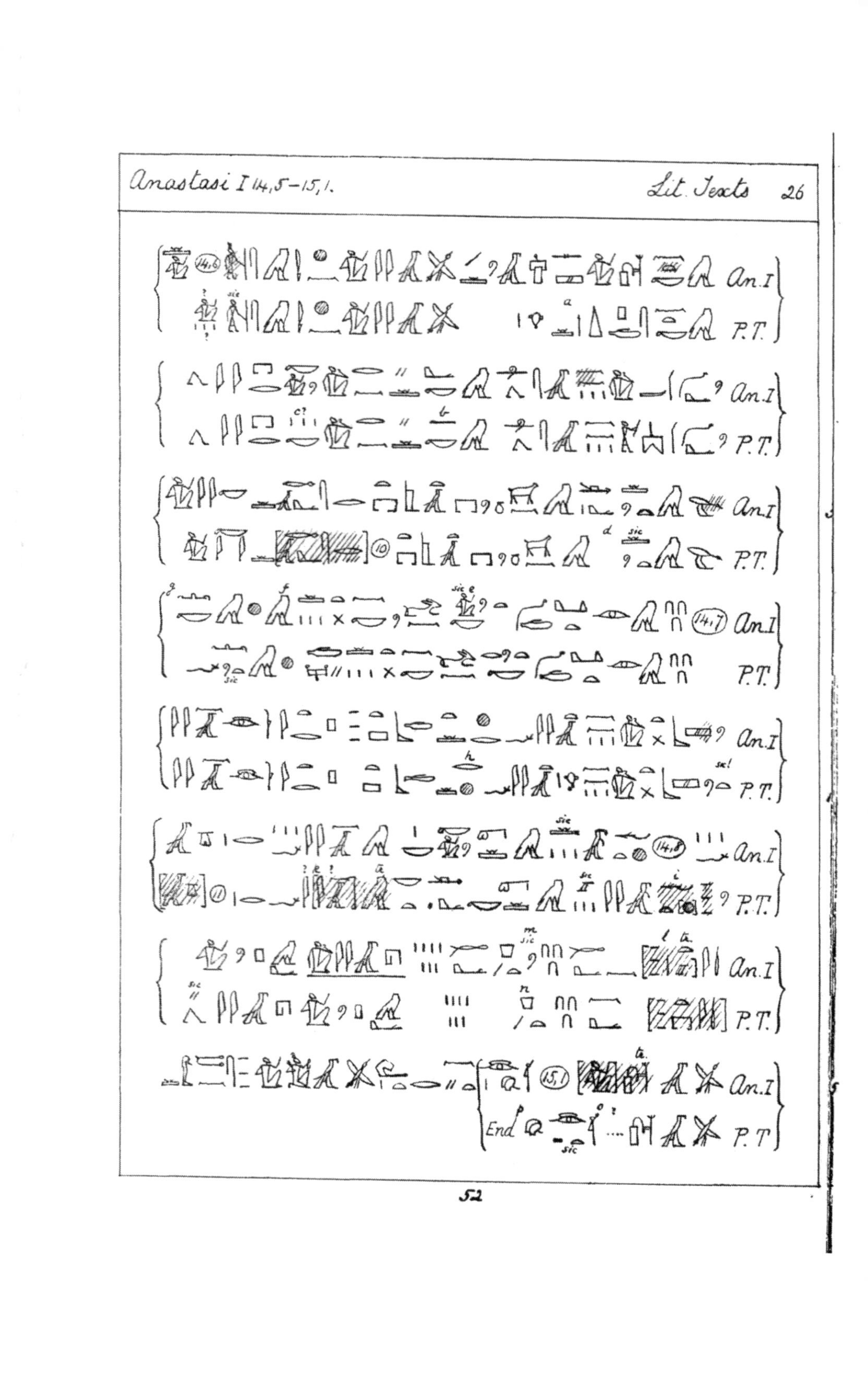

An. I
P.T.

An. I
P.T.

An. I
P.T.

An. I
P.T.

An. I
P.T.

An. I
P.T.

An. I
P.T.

An. I
End P.T.

2.[a] Indeterminate shape, almost like [hieroglyph], see facs.; clearly derived from [hieroglyph].

4.[b] Like hieratic [hieroglyph], a corruption of [hieroglyph]; cf. p. 24a, note 14.[l] — [c] Indefinite sign, see facs.

6.[d] Wr wrongly omitted.

7.[e] Emend [hieroglyph] into [hieroglyph], cf. P. T. — [f] Superfluous m. — [g] Here a misplaced fragment above [line.

10.[h] So too above p. 25, l. 12.

12.[i] Obscure traces, probably to be interpreted thus. — [k] The stroke after the lacuna is more like [hieroglyph]; and the lacuna is small for [n3i]i.

13.[l] The word, cf. 14,3 above. — [m] Cf. 16,7, a spelling intermediate between [hieroglyph] ([hieroglyph]) and [hieroglyph] ([hieroglyph]); on the latter see p. 25[a], note 4[c].

14.[n] Added above the line.

16.[o] I saw [hieroglyph], a corruption of [hieroglyph]? — [p] P. T. ends abruptly with a short line.

[hieroglyphs]

[hieroglyphs] ā. a [hieroglyphs] (15,2) [hieroglyphs] ā. [hieroglyphs] ā. [hieroglyphs]

sic [hieroglyphs]

[hieroglyphs] sic [hieroglyphs] sic [hieroglyphs] (15,3) [hieroglyphs]

[hieroglyphs] b [hieroglyphs] 5

[hieroglyphs] c [hieroglyphs] (15,4) [hieroglyphs]

[hieroglyphs] sic d [hieroglyphs]

[hieroglyphs] (15,5) [hieroglyphs]

[hieroglyphs] (15,6) [hieroglyphs] sic [hieroglyphs] ā. [hieroglyphs]

[hieroglyphs] ā. [hieroglyphs] ? ā. [hieroglyphs] e [hieroglyphs] 10

[hieroglyphs] ā. [hieroglyphs] (15,7) [hieroglyphs]

[hieroglyphs]

[hieroglyphs] (15,8) [hieroglyphs] ā. [hieroglyphs] ā. [hieroglyphs]

[hieroglyphs] g [hieroglyphs] f [hieroglyphs]

[hieroglyphs] (16,1) [hieroglyphs] 15

[hieroglyphs] ā. [hieroglyphs]

[hieroglyphs] i [hieroglyphs] (16,2) [hieroglyphs] h [hieroglyphs]

l k [hieroglyphs]

2.[a] The dash of [hieroglyph] is visible in a trace.

5.[b] For this curiously written expression compare the facsimile.

6.[c] Or possibly [hieroglyph].

7.[d] The Ms. has a clear [hieroglyph], not (e.g.) [hieroglyph].

10.[e] Original properly [hieroglyph], as in inḥ, below 19,4; conversely see above p. 2a, note 13.[i]

14.[f] [hieroglyph] made large like [hieroglyph], as often below or above long signs. — [g] Not room for [hieroglyph].

17.[h] There seems just enough room for a short low sign before sp. — [i] [hieroglyph] is superfluous.

18.[k] Original [hieroglyph] sic. — [l] By error for [hieroglyph].

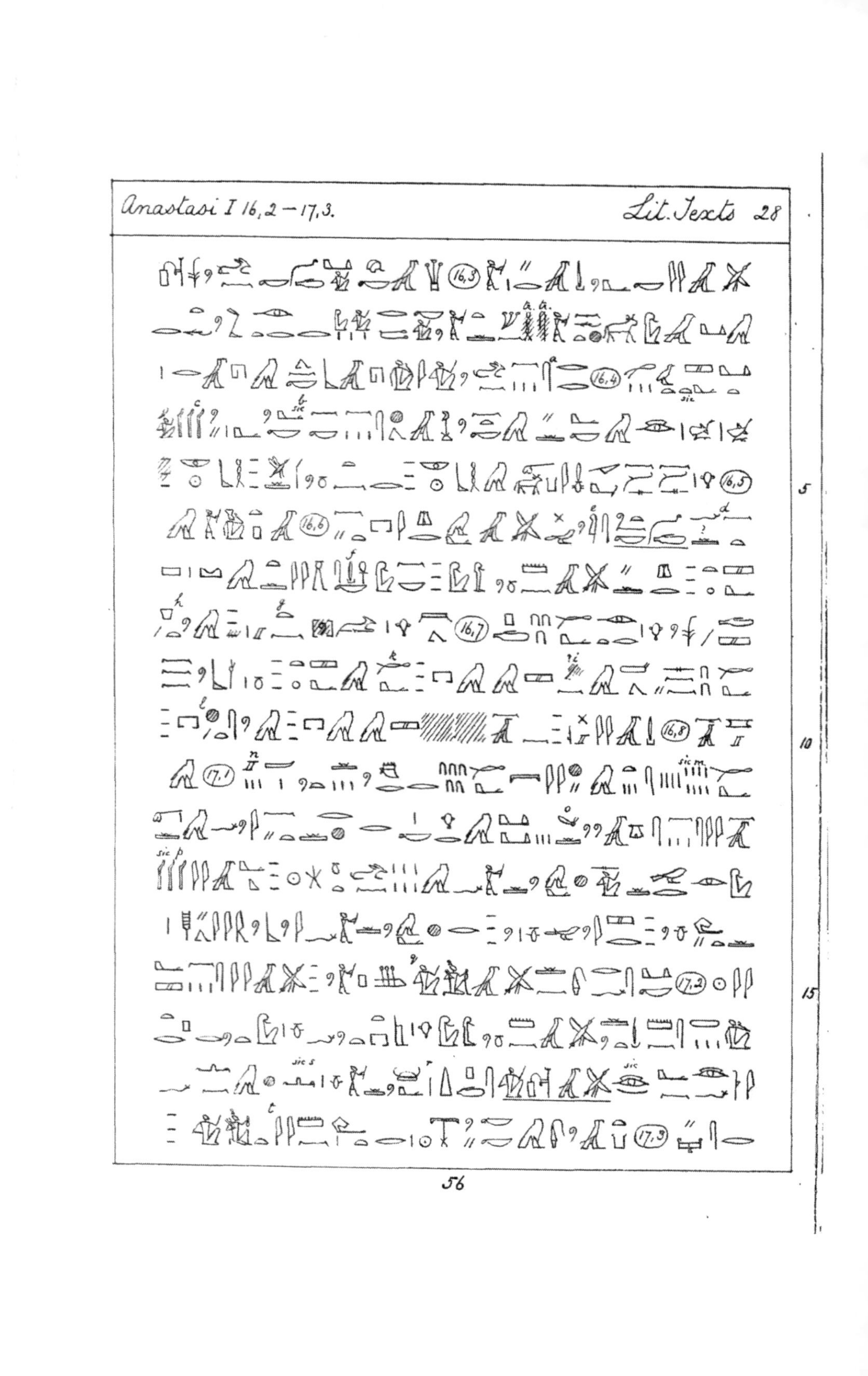

(16,3) (16,4)

(16,5) 5

(16,6)

(16,7)

(16,8) 10

(17,1)

(17,2) 15

(17,3)

3.[a] Emend [hieroglyphs].

4.[b] Read [hieroglyphs]. — [c] Here there has been an obscure corruption.

6.[d] [hieroglyphs]; evidently a corruption, perhaps of [hieroglyphs]. — [e] So wrongly for [hieroglyph].

7.[f] Cf. above 14,1.

8.[g] [hieroglyph] is made like [hieroglyph] above the long sign [hieroglyph], as often in this word. — [h] See above p. 26a, note 13.[m]

9.[i] See the facsimile, which is exact; perhaps [hieroglyphs]? — [k] For [hieroglyphs], see p. 25a, notes 5.[f] 15.[o]

10.[l] The determinatives [hieroglyphs] are wrong.

11.[m] Original [hieroglyphs], probably to be emended into [hieroglyphs] 40. — [n] Or [hieroglyphs]?

12.[o] Or [hieroglyphs]?

13.[p] Mistake for [hieroglyph].

15.[q] [hieroglyph] wrongly omitted.

17.[r] Emend *spd* [hieroglyphs], cf. p. 26, l. 2. — [s] Read [hieroglyphs], cf. 1,1.

18.[t] Emend *mnfy-t*; the same corruption above 14,2.

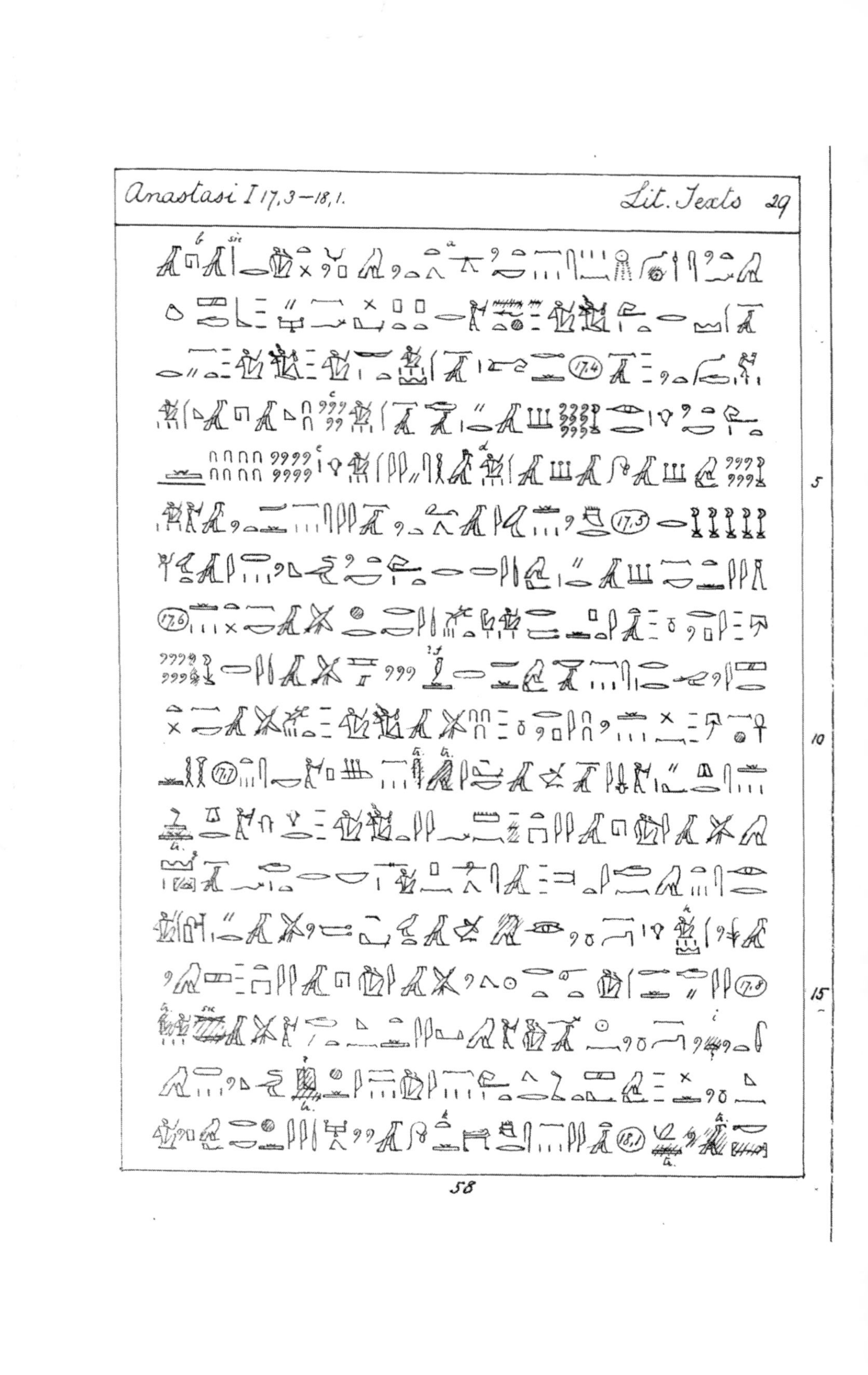

(17,4)

5

(17,5)

(17,6)

10

(17,7)

15

(17,8)

(18,1)

1.[a] Read [hieroglyphs]. — [b] [hieroglyphs] must be a corruption of [hieroglyphs]; emend therefore [hieroglyphs]; the context clearly cannot refer to a quarry like Hammamat (R-h-n).

4.[c] See facsimile; the Ms. apparently had originally [hieratic numeral], later corrected into [hieratic numeral]. The latter figure suits the total (1900 + 620 + 1600 + 880 = 5000), but it seems likely that a number has fallen out after M-š-w-š.

5.[d] A number is probably lost, see last note. — [e] Either omit [hieroglyphs] or insert [hieroglyphs] after it.

9.[f] See the facsimile for this problematical group; so too in [hieroglyphs] [Anast. IV 13, 12.

13.[g]–14.[h] Emend [hieroglyphs].

16.[i] If this be the correct reading the second [hieroglyphs] is crowded, which might account for the abnormal form of [hieroglyph]. The traces suggest [hieroglyphs] (by error for [hieroglyphs] ??) more than anything else; or can the true reading be [hieroglyphs]?

18.[k] Or [hieroglyphs] (??); one would however then expect [hieroglyphs].

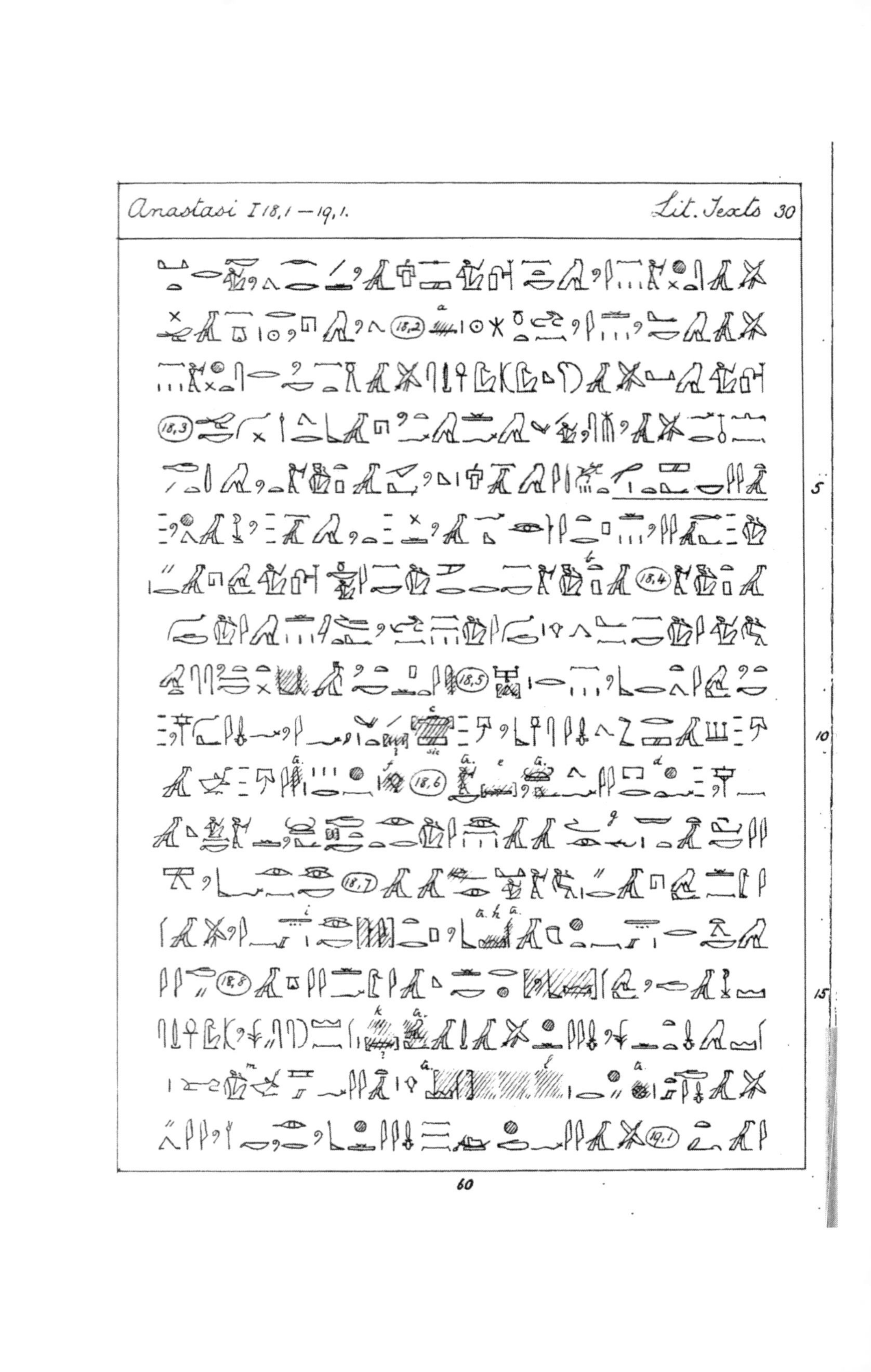

(18,2) (18,3) (18,4) (18,5) (18,6) (18,7) (18,8) (19,1)

5

10

15

2.[a] Probably a dash serving to complete the line (Füllstrich).

7.[b] Probably a dittograph owing to change of line.

10.[c] For the omission of r cf. d3r 15,6; there is a trace under / .

11.[d] Ms. gives . — [e] See facsimile; the reading wḫ' is certain. — [f] Ḥr superfluous, as [often.

12.[g] Above a spot of red ink; since another such, but smaller, occurs above , this is probably fortuitous, not an isolated verse-point.

14.[h] Above a misplaced fragment which is difficult to locate; but cf. , the true place of below note 17[l]. — [i] A deleted sign adjoins the stroke | ; cf. facs..

16.[k] So too Max Müller, O.L.Z. I 382; is almost certain; in place of | it would be possible to read | , but this latter hardly fills the lacuna.

17.[m] After the stroke | is omitted.

17.[l] After | small and puzzling traces are visible, not suiting either or ; the usual reading Ḥ-r.t is wrong. Possibly the misplaced fragment of note 14[h] belongs here, in which case the true reading might be .

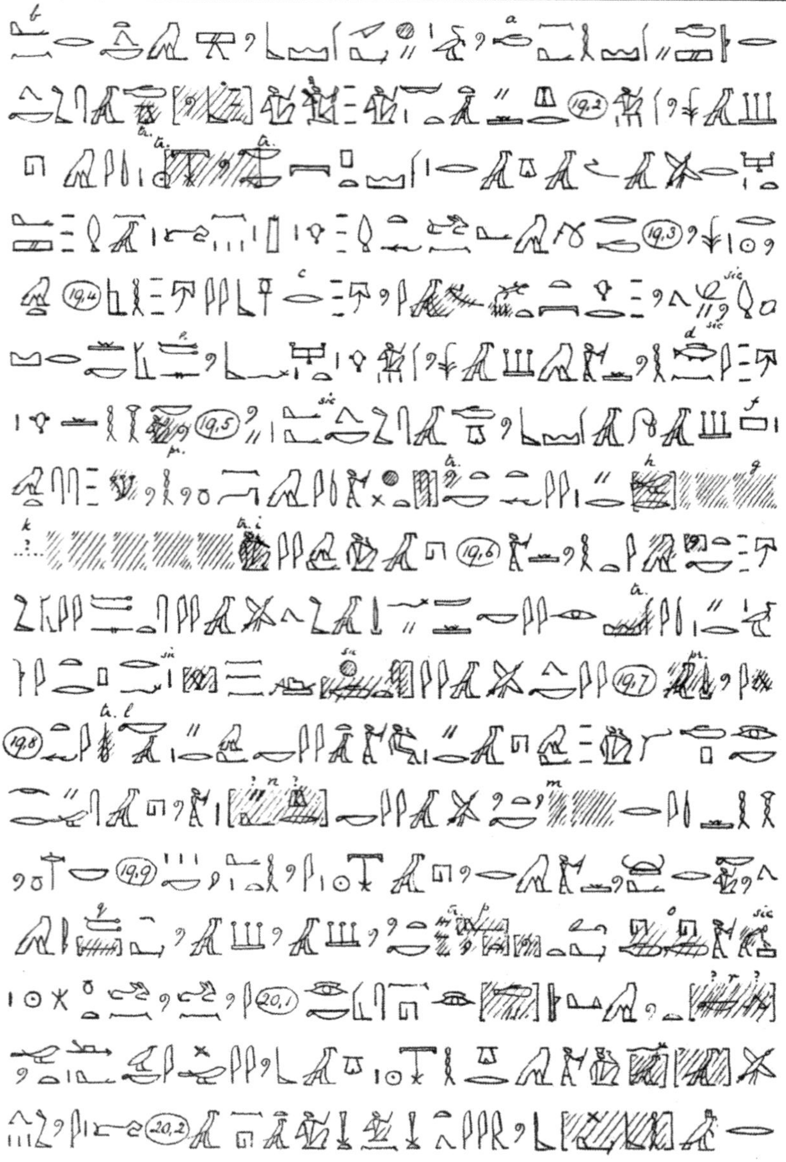

(19,2) (19,3) (19,4) (19,5) (19,6) (19,7) (19,8) (19,9) (20,1) (20,2)

a b c d f g h k l m n p q r; sic; tr.; pr.

1.[a] Small, but not like [hieroglyph]. — [b] Emend [hieroglyph].

5.[c] Like [hieroglyph]; however [hieroglyph] is certainly to be read.

6.[d] Really [hieroglyph] in the Ms., as in sin 15,6 etc. — [e] Ms. [hieroglyph] · in the preceding clause read [hieroglyph].

7.[f] So wrongly for [hieroglyph]; contrast 21,5.6.

8.[g] For the traces after hr see the facsimile. — [h] Cf. 24,4.

9.[i] A trace high up, which cannot belong to [hieroglyph]. — [k] [hieroglyph]

12.[l] Emend m-r-k-?[hieroglyph]-t.

13.[m] For the traces see facsimile. — [n] Elsewhere we find p-k m-r-ʿ (20,3); this can however not be emended here.

15.[o] See facsimile, now partially destroyed; hdhd is a more probable reading than hnhn. — [p] Ks "bones" is not a possible restoration, because -tsk indicates a feminine word.

15.[q] Sic; but the verb should be spelt with [hieroglyph], not [hieroglyph], cf. 16,2.

16.[r] [hieroglyph] alone would not suffice to fill the lacuna. — From a final examination of this passage it seems possible that the preceding signs stand on a misplaced fragment of papyrus.

[illegible] sic [illegible] a [illegible] a. [illegible]
[illegible] (20,3) [illegible] pl. a [illegible]
[illegible]
[illegible] (20,4) [illegible] a. b [illegible]
[illegible] c [illegible] 5
[illegible] (20,5) [illegible]
[illegible] d [illegible]
[illegible] (20,6) [illegible]
[illegible] a. [illegible]
[illegible] (20,7) [illegible] sic [illegible] 10
[illegible] e [illegible]
[illegible] a. [illegible] (20,8) [illegible]
[illegible] a. [illegible]
[illegible] f [illegible] (21,1) [illegible]
[illegible] 15
[illegible] (21,2) [illegible]
[illegible] g [illegible]
[illegible] (21,3) [illegible]

2.[a] Traces of a vertical sign, not 𓋹.

4.[b] There is nothing more lost.

5.[c] An almost certain trace of the tail of m.

7.[d] Erroneously written for [hieroglyphs].

11.[e] Cf. below 21,1.

14.[f] Not room for [hieroglyph]; n3 is therefore certain.

17.[g] Read [hieroglyph]

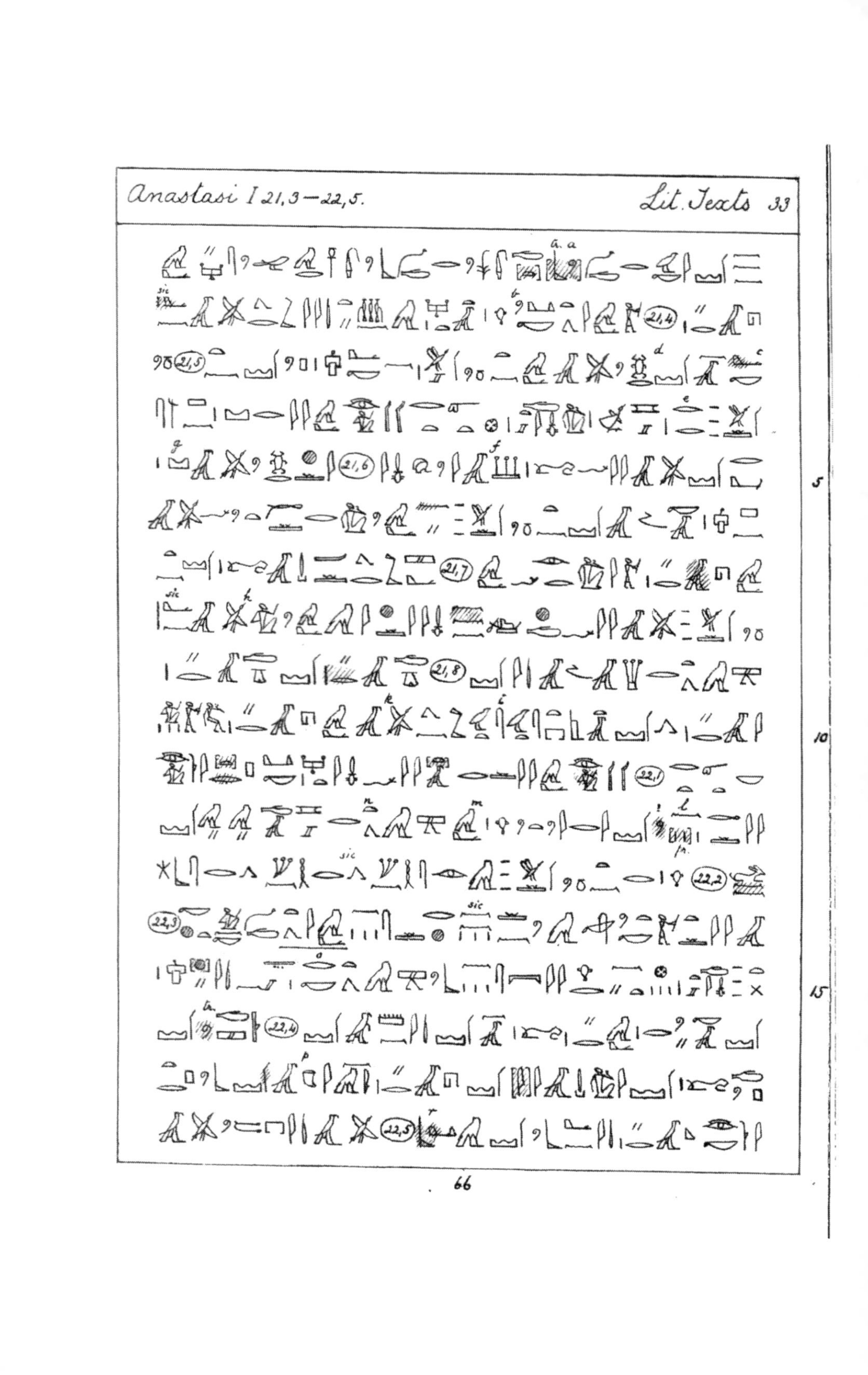

(21,4) (21,5)

5

(21,6) (21,7) (21,8)

10

(22,1) (22,2) (22,3)

15

(22,4) (22,5)

1.[a] The misplaced fragment at the beginning of the line, consisting of followed by the trace of a vertical stroke, very possibly belongs to wbd here.

2.[b] Emend (?), cf. imi-wi 21,7.

3.[c-d] is nearly, and the preceding quite, certain. We should probably read Burchardt no. 288.

4.[e] Perhaps emend for .

5.[f] Burchardt (no. 636) rightly transposes and reads . —[g] Ms. .

8.[h] Insert after imi-wi (?).

10.[i] Swtwt; the second s is superfluous. —[k] P3 must be omitted because of nb.

12.[l] ; the final w is very unlikely, but an alternative hard to find. —[m-n] Read as in 21,6-7.

15.[o] Wrongly for ; the mistake is perhaps due to being inserted above the line in the manuscript from which the scribe of Anastasi I copied.

17.[p] is derived from a misunderstood hieratic (det. of nmi), which the scribe understood as t3; hence the following 3. Cf. the spelling for in Leiden [Admonitions.

18.[r] The upper portion of is displaced in original and facsimile.

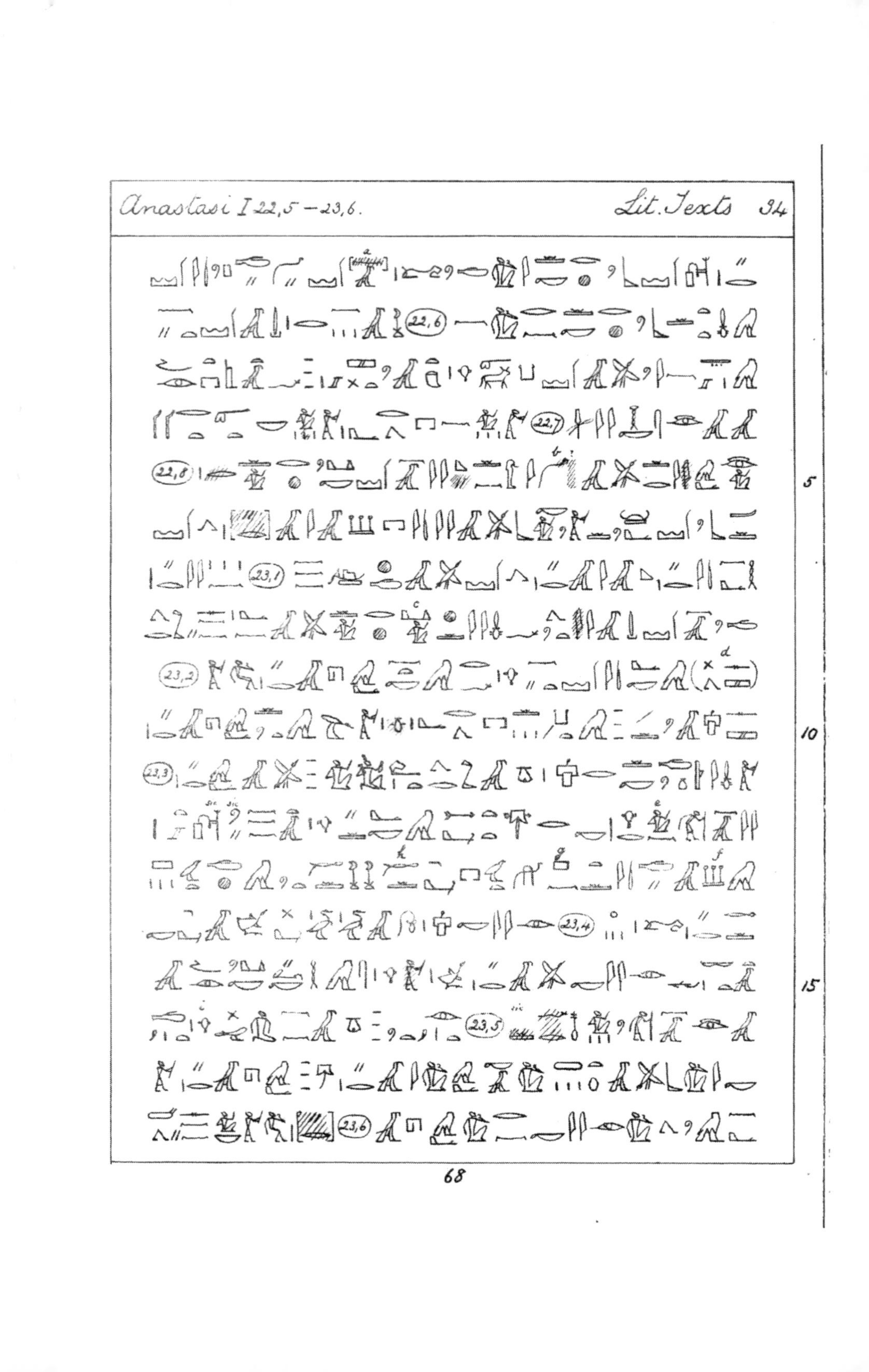

(22,6)

(22,7)

(22,8) 5

(23,1)

(23,2)

10

(23,3)

(23,4)

15

(23,5)

(23,6)

1.[a] [hieroglyphs] would also be a possible reading.

5.[b] [hieroglyphs, erased] ; the upright sign, which is a correction, is not [hieroglyph] nor [hieroglyph] nor [hieroglyph].

8.[c] Emend [hieroglyphs], cf. 22,7 end.

9.[d] Ms. thus :– [hieroglyphs] ; it is doubtful whether [hieroglyphs] is a gloss on, or correction of, snni.

12.[e] [hieroglyph] is a correction over an erased [hieroglyph] (?).

13.[f].[g] Emend [hieroglyphs] cf. 24,3. – [h] Read [hieroglyphs] ; for the converse confusion, cf. 14,5.

16.[i] Hr. probably superfluous; cf. p. 13a, note 7[f]; p. 30a, note 11[f].

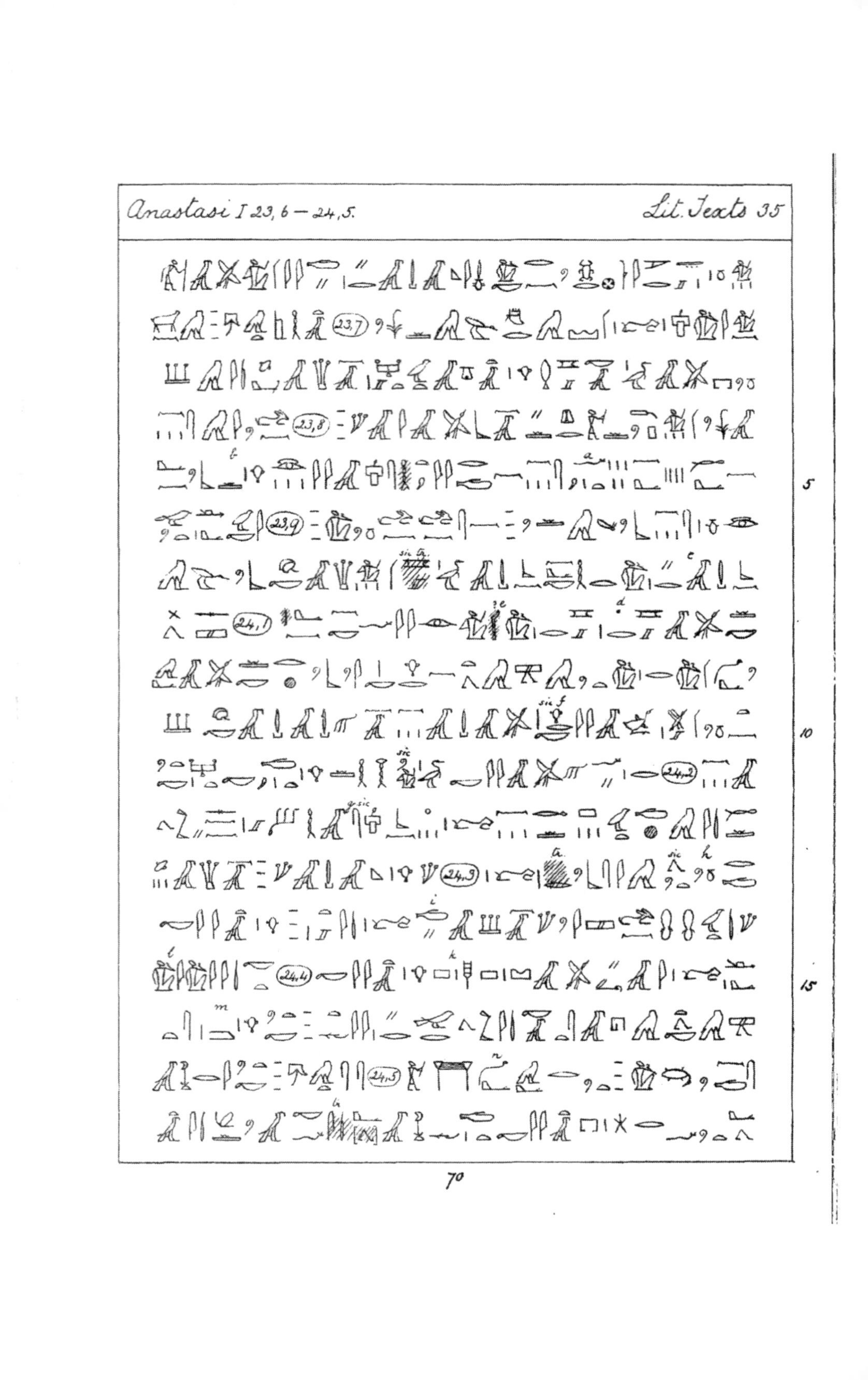

(23,7) (23,8) (23,9) (24,1) (24,2) (24,3) (24,4) (24,5)

5

10

15

5.[a] A corrupt word; perhaps read [hieroglyphs] (?). — [b] [hieroglyph], read [hieroglyph]; cf. p. 11a, note 5[g] (?).

7.[c] Probably emend [hieroglyphs] cf. 16,2.

8.[d] A tiny dot, high up; probably fortuitous. — [e] Probably so, and certainly not [hieroglyph].

10.[f] Sic; a corruption of [hieroglyph].

12.[g] Read [hieroglyphs]; the two signs resemble one another closely, whence the transposition.

13.[h] [hieroglyphs], a transparent corruption of [hieroglyphs].

14.[i] The [hieroglyph] is made quite small, almost like [hieroglyph], above the small sign [hieroglyph].

15.[k] Emend [hieroglyphs] (?). [l] Dittograph.

16.[m] [hieroglyphs] is corrected out of [hieroglyphs].

17.[n] Emend [hieroglyphs].

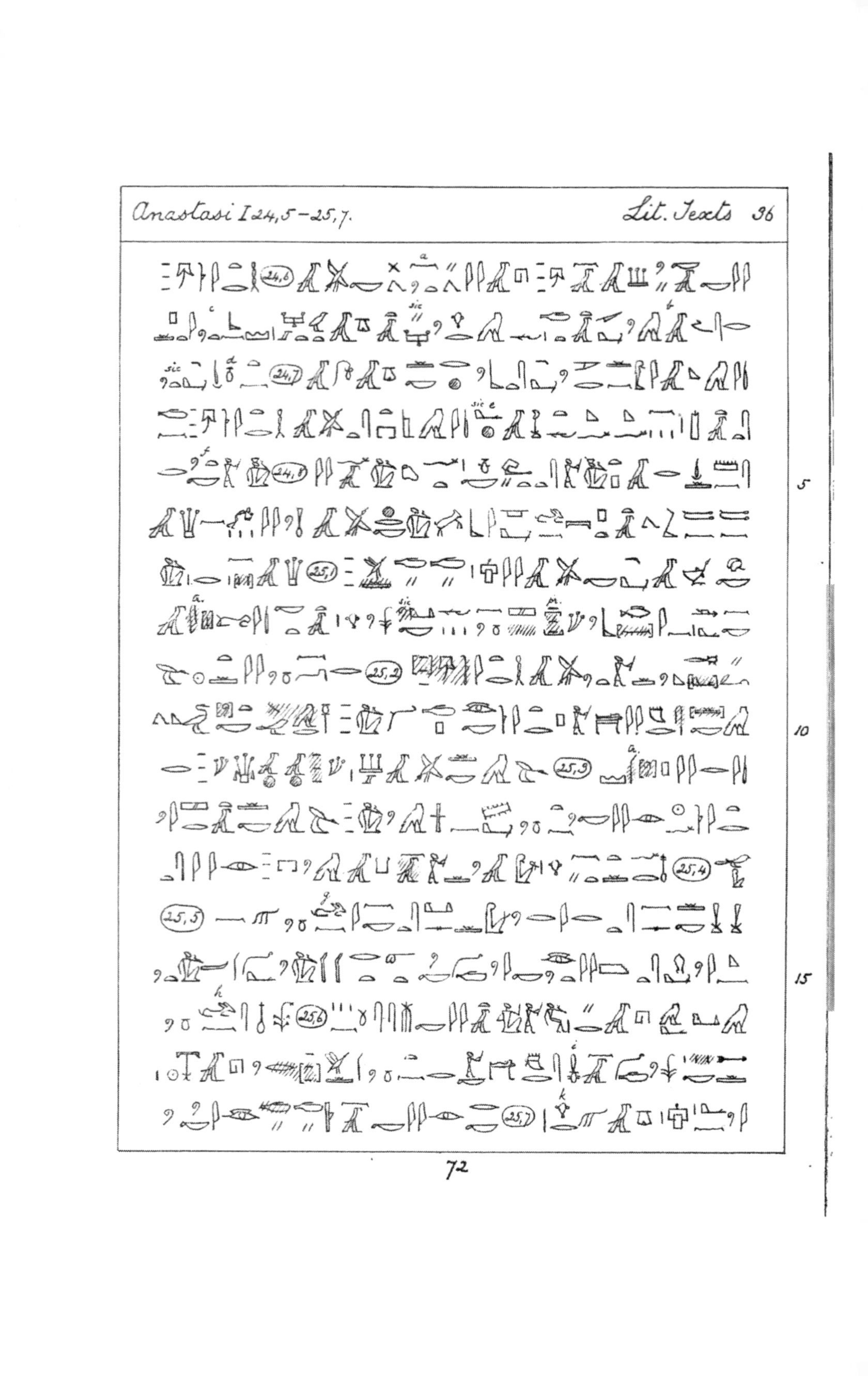

(24,6) a

sic c b

sic d (24,7)

sic e

f (24,8) 5

(25,1)

a sic m

(25,2)

10

(25,3) a

(25,4)

(25,5) g

15

h (25,6)

i

k (25,7)

1.[a] Emend [hieroglyphs]; the corruption of [hieroglyph] into [hieroglyph] is not rare in hieratic.

2.[b] Emend [hieroglyphs] (?). — [c] Read [hieroglyphs] for [hieroglyphs].

3.[d] Read g-w-t-n [hieroglyphs] cf. Koller 3, 2 = Anast. IV 3, 1.

4.[e] Read [hieroglyphs].

5.[f] Read [hieroglyphs] cf. p. 21a, note 3[e]; p. 32a, note 7[d].

14.[g] Ms. has [hieroglyphs] for [hieroglyphs] cf. sun 25,6; p. 2a, note 13[i]; etc.

16.[h] See note 14[g].

17.[i] Without [hieroglyph], evidently corrupt.

18.[k] A ligature as in 28, 4.

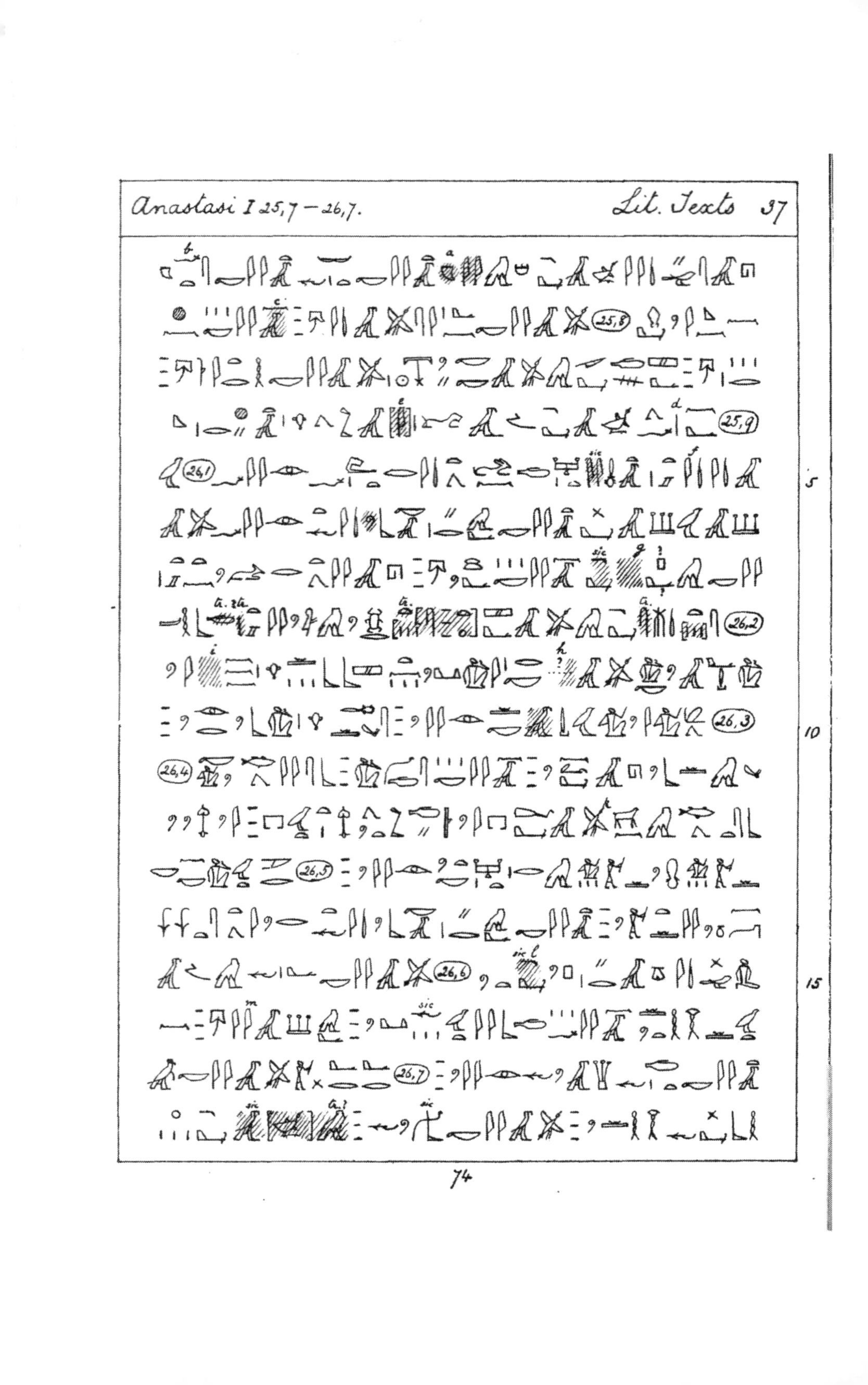

(25,8)

(25,9)

(26,1) 5

(26,2)

(26,3) 10

(26,4)

(26,5)

(26,6) 15

(26,7)

1.[a] [hieroglyph] ; possibly [hieroglyph], but [hieroglyph] has a different form e.g. after sft below. Or perhaps [hieroglyph]? — [b] [hieroglyph] like [hieroglyph] as practically always in this word.

2.[c] The papyrus is unduly squeezed together at this point.

4.[d] Corrupted out of [hieroglyph]. — [e] See note 2[c] above.

5.[f] Dittograph.

7.[g] [hieroglyph] ; the hieratic signs may well be corruptions.

9.[h] The traces (see facs.) look rather like the left end of [hieroglyph]; this would however not suit the context. — [i] [hieroglyph]

12.[k] Emend [hieroglyph].

15.[l] Certainly so, not [hieroglyph].

16.[m] The first [hieroglyph] has a small meaningless appendage at the bottom, see facsimile.

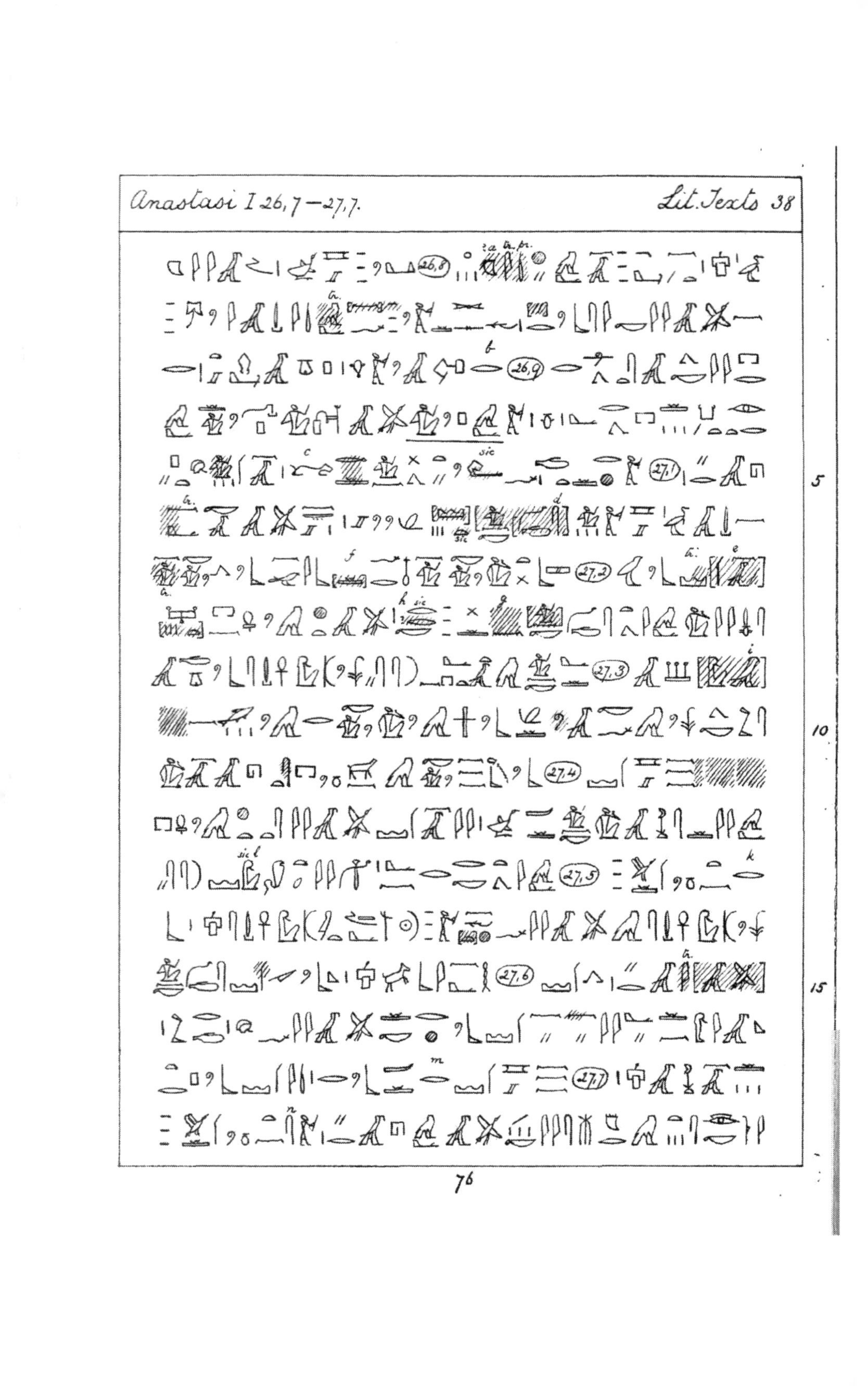

5

10

15

1.[a] An upright sign; might possibly be [hieroglyph] or [hieroglyph].

3.[b] Dittograph.

5.[c] Emend [hieroglyphs] as in 17,3.

6.[d] There appears to be room for s<u>dd</u>.

7.[e] A misplaced fragment covers the right half of [hieroglyph]. — [f] Only a small sign lost.

8.[g] Perhaps [hieroglyphs] by error for [hieroglyphs]. — [h] Probable; certainly neither [hieroglyph] nor [hieroglyph].

9.[i] To left of [hieroglyph] a slanting trace (see facs.) which might possibly belong to [hieroglyph].

13.[k] Omit [hieroglyph]. — [l] [hieroglyph] is a corruption of [hieroglyph].

17.[m] This [hieroglyph] is probably superfluous.

18.[n] For [hieroglyph] emend [hieroglyphs] or [hieroglyphs].

(27,8)

(27,9)

(28,1)

5

(28,2)

(28,3)

10

(28,4)

15

(28,5)

1.[a] [hieroglyph] is corrected out of a tall erased sign; stt is probably a corruption of [hieroglyphs]

2.[b] For [hieroglyphs] emend [hieroglyphs].

5.[c] The hinder arm of this sign has been omitted.

6.[d] The Ms. reading is perhaps a blending of two different readings [hieroglyphs] and [hieroglyphs].

11.[e] Emend [hieroglyphs], cf. 10,3.

14.[f] An obscure corruption.

15.[g] [hieroglyph] should probably be omitted.

17.[h] Ms. [hieroglyphs]. – [i] [hieroglyph] is superfluous, cf. 6,7.

18.[k] Read [hieroglyphs].

[hieroglyphic text] [a] (28,6) [hieroglyphic text] [a]

[hieroglyphic text] [a] [hieroglyphic text] [a] [a] (28,7) [a] [hieroglyphic text] 5

[hieroglyphic text] [b] [hieroglyphic text]

(28,8) [hieroglyphic text] [c] [hieroglyphic text] [d] [hieroglyphic text]

End ? [f] [hieroglyphic text] [e] [hieroglyphic text] 10

3.[a] For this sign see Möller, Hieratische Paläographie II, no. 505 footnote.

6.[b] [hieroglyph] is here made like [hieroglyph]; [hieroglyphs] is meant.

7.[c] A corruption of [hieroglyphs] as in Sinuhe B.

8.[d] Emend [hieroglyphs].

10.[e] Not [hieroglyphs]. — [f] Here a fragment bearing a quite illegible sign, probably misplaced. The final sign might also be [hieroglyph]; if wꜣ "to be far" is meant, we should expect as determinatives [hieroglyphs] not [hieroglyph] (sic) [[hieroglyph]].

2. A collection of model letters,

transcribed from Pap. Koller (Pap. Berlin 3043) and certain other Mss.

a. The equipment of a Syrian expedition.

1.[a] Before 1.1 an unknown number of pages are lost.

3.[b] Or large ⌓? Or else a corruption of [hieroglyph] kp?

5.[c] [hieroglyphs]; if the reading [hieroglyph] is correct, that sign is here different from the forms elsewhere on this page.

6.[d] [hieroglyph] small like ⌓. — [e] The plural strokes are omitted.

7.[f] [hieroglyphs]; the sign before I [hieroglyph] may be the remains of an erroneous [hieroglyph].

8.[g] [hieroglyph] unlike the other examples of [hieroglyph] on this page.

9.[h] Made like [hieroglyph], see above p. 37a, note 1.[b]

10.[i] Emend [hieroglyphs] as below 2,1.

11.[k] Emend <u>m·r-k-<9[hieroglyph]>-t</u>, as in <u>Anast.</u> I 19,7 (see above p. 31a, note 12.[l]).

12.[l] Dittography.

13.[m] Probably so to be read, but written [hieroglyph]. cf. above note 9.[h]

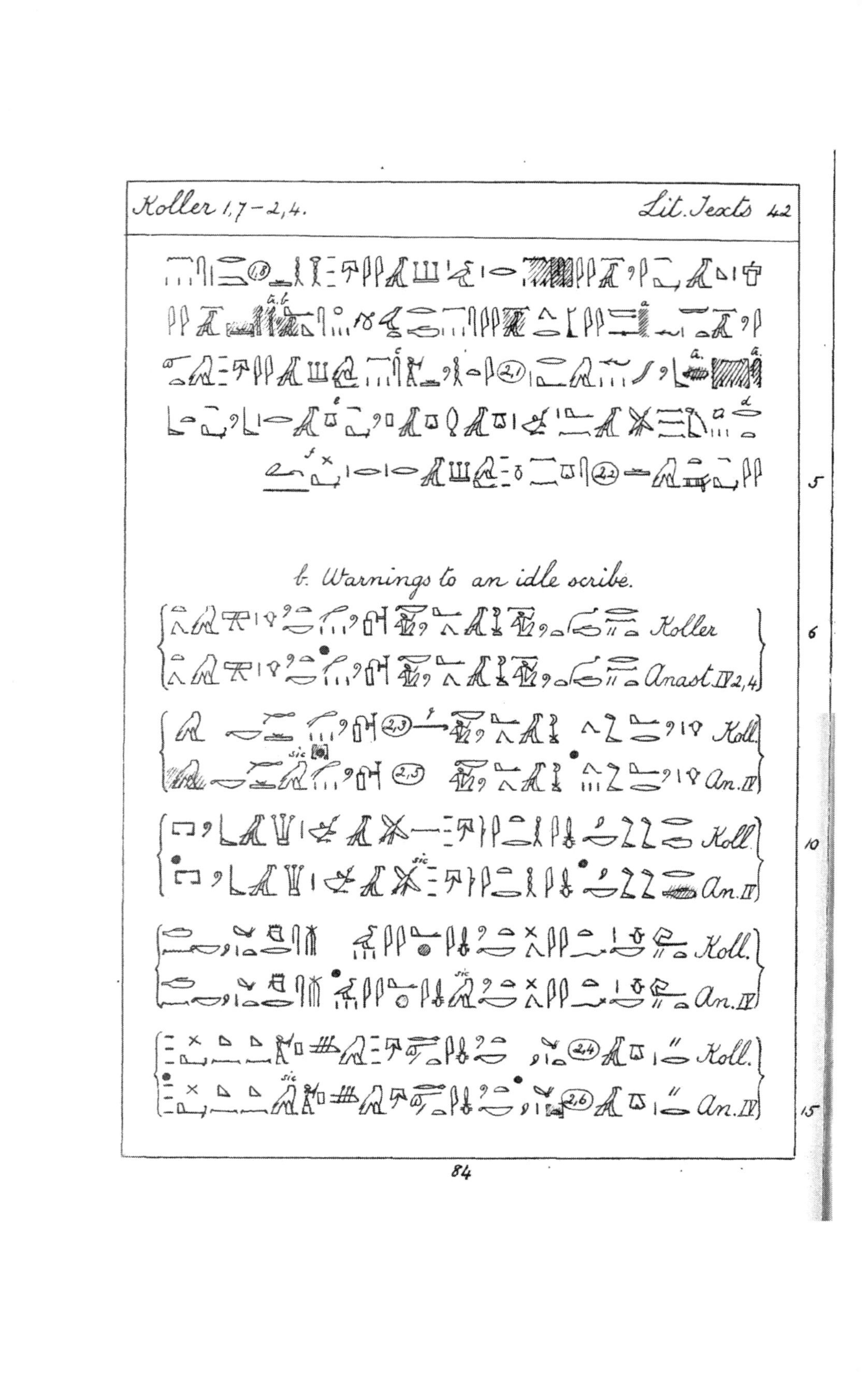

5

b. Warnings to an idle scribe.

Koller 6

Anast. IV 2,4

Koll.

sic

An. IV

Koll. 10

sic

An. IV

Koll.

sic

An. IV

sic

Koll.

An. IV 15

2.[a] Low down a trace, possibly a small [hieroglyph]; anyhow not a part of [hieroglyph]. — [b] Wrongly mounted; only a tiny trace of the first [hieroglyph] remains.

3.[c] For -sn emend [hieroglyphs] ?

4.[d] Like [hieroglyph].

4.[e]-5.[f] Anast. IV 16, 11-12 in a similar context:- [hieroglyphs]

8.[g] A meaningless stroke (Füllstrich).

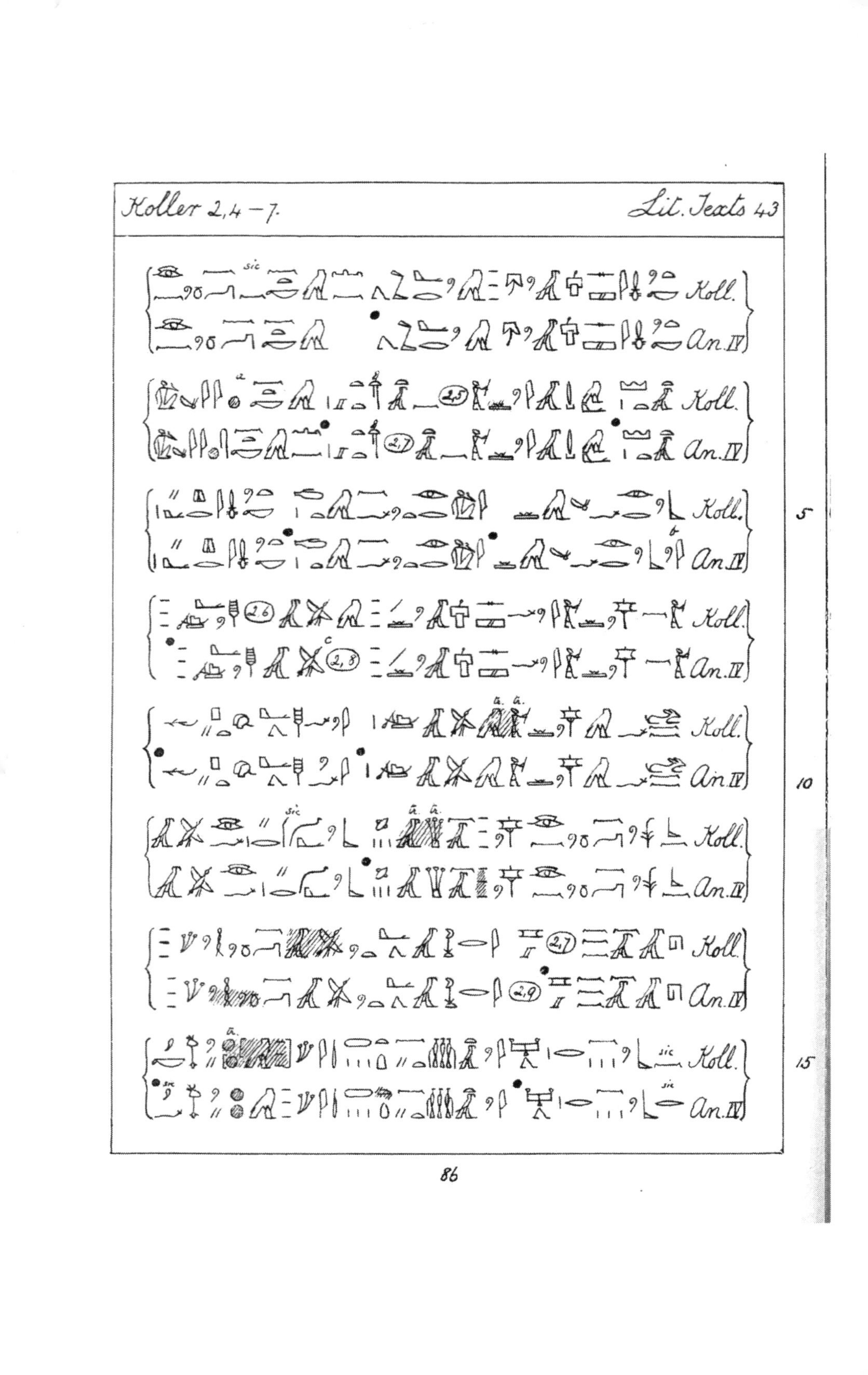

sic Koll.
An. IV

a Koll.
An. IV

Koll. 5
b An. IV

Koll.
c An. IV

a. a. Koll.
An. IV 10

sic a. a. Koll.
An. IV

Koll.
An. IV

a. sic Koll. 15
sic sic An. IV

3.[a] There is no trace of an [hieroglyph], nor has this sign been written and for some reason deleted.

6.[b] _Iw_ is lacking not only in _Koller_, but in the similar passage _Anast._ I 6, 6.

8.[c] There is no [hieroglyph] before _p3_, though there is a fortuitous black speck.

(2,8) Koll.
(2,10) An. IV

Koll.
An. IV
(5,1) Anastasi V

5

sic b
Koll.
An. IV
End c ... An. V

pr. e (2,9) d Koll.
(2,11) An. IV

10

pr. h
Koll.
i
An. IV

k j (3,1) Koll.
(2,12) An. IV

15

Koll.
An. IV

3.[a] After ḏ(3)p(w) the traces [hieroglyph] are still visible.

6.[b] The lacuna is too large for iwf only.

8.[c] The top line of the page alone is preserved, and this only in part. However see below note 13.[l]

9.[d] Small like [hieroglyph], cf. 2,1. — [e] After ii one or more signs entirely washed out, save for a horizontal stroke at bottom; [hieroglyph] is nearly certain

10.[f] At top a trace which suits [hieroglyph]. — [g] [hieroglyph]; the reading is far from certain.

11.[h] [hieroglyph]

12.[i] [hieroglyph]; these traces do not suit the reading n mḥ 7 very well.

13.[l] The word [hieroglyph] written as a correction above Anast. V 5,1 probably belongs here. — [k] Only one tall sign, perhaps [[hieroglyph]].

Koll.
An. IV

Koll.
An. IV

Koll. 5
An. IV

Koll.
An. IV

Koll.
An. IV 10

Koll.
An. IV

c. A letter concerning Nubian tribute.

Koller
Spieg., Hier. Ostr. 39 14

1.[a] [hieratic sign] ; contrast [sign] in _dbw_ 1,8. — [b] Here a deleted [sign], which apparently does not belong to the earlier text.

2.[c] [hieratic sign] rather like the det. of _db_ in _Anast._ IV 17,8.

11.[d] Perhaps rightly interpreted by Möller (_Paläographie_ II no. 101 note) as a contraction of [sign] "chapter" "section"; but certainly understood by the N.K. scribe as [sign].

14.[e–f] For the probable size of the lacuna, see lines 4.5.

Koll.

Sp., H.O.

Koll.

Sp., H.O.

Koll. 5

Sp., H.O.

Koll.

Sp., H.O.

Koll.

Sp., H.O. 10

Koll.

Sp., H.O.

Koll.

Sp., H.O.

Koll. 15

End

Sp., H.O.

1.[a] For [hieroglyphs] emend [hieroglyphs].

4.[b] The ostr. had clearly a version quite different from that of Koller; the lacuna is probably a little shorter than that of ll. 4.5. After the lacuna Spiegelberg reads [hieroglyphs]; but the word for "garrison" is iwꜥyt not iwyt.

6.[c–d] Or perhaps better [hieroglyphs], cf. Koller 1,1; 5,5.

9.[e] Probably no mere "Füllstrich". —[f] Badly mounted; not ng.

12.[g] Ostr. had clearly a longer list of animals; for the length of lacuna cf. ll. 4.5.

13.[h] The hieratic sign serves at once for [hieroglyph] and for [hieroglyph]; see above p. 25a, note 4[c]. —[i] Emend [hieroglyph].

16.[k–l] Spieg.'s facs. gives [hieroglyphs] after the long lacuna at the beginning of l. 7. Impossible to connect with text of Koller. —[m] As colophon (l. 8) in red, [hieroglyphs] ......

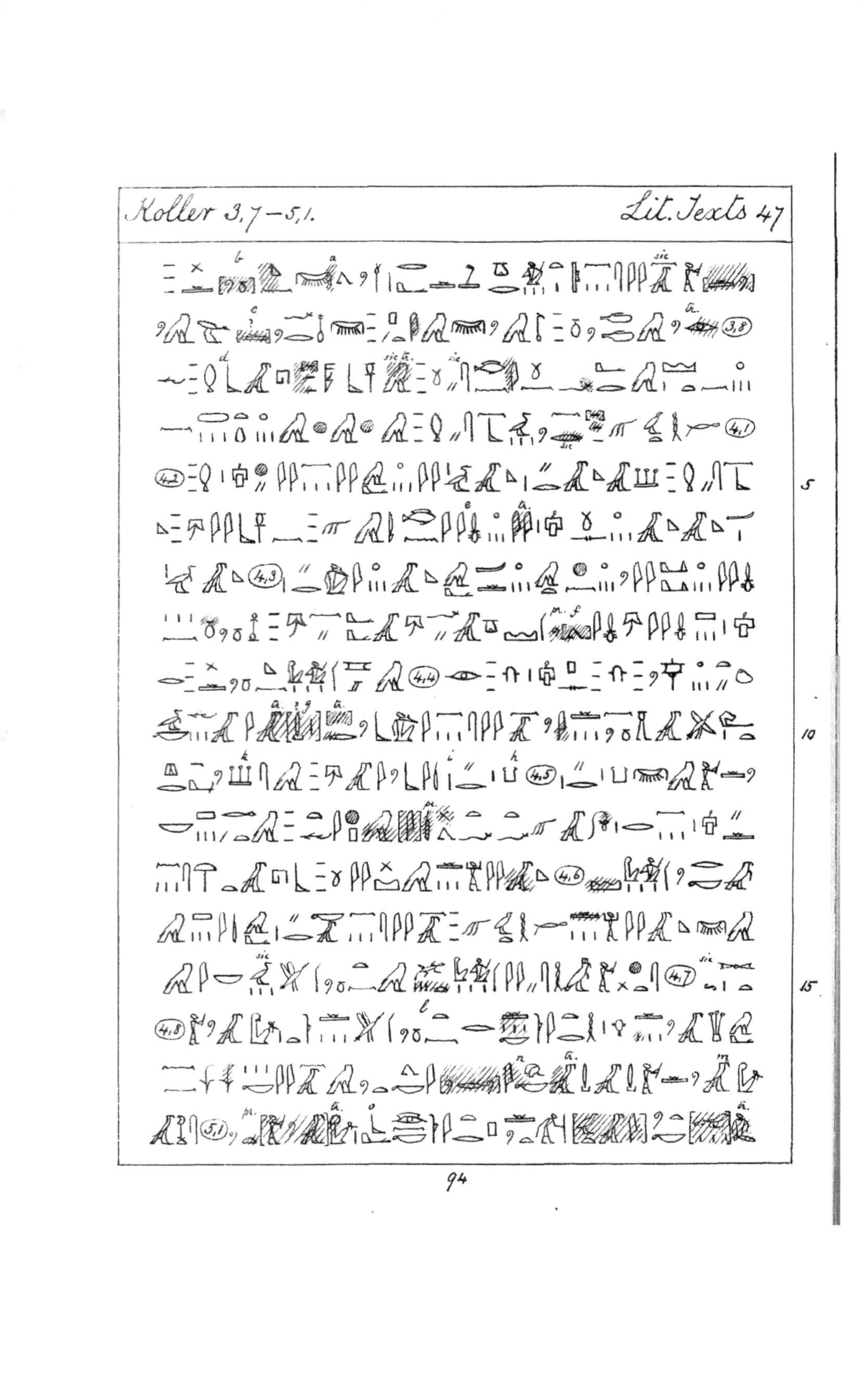

5

10

15

1.[a] The signs [hieroglyphs] are on a displaced fragment; possibly there was an m between [hieroglyph] and [hieroglyph]. — [b] Perhaps hardly room for [[hieroglyphs]].

2.[c] Only a small sign lost.

3.[d] Emend [hieroglyphs].

6.[e] [hieroglyphs] appears to be superfluous.

8.[f] [hieroglyphs]

10.[g] The scanty space hardly admits of any other restoration.

11.[h-i] Perhaps dittography. — [k] [hieroglyphs] not improbably a corrupt group.

16.[l] This word is a correction; the [hieroglyph] is above a deleted [hieroglyph].

17.[m] Dittograph. — [n] [hieroglyphs] on a displaced fragment.

18.[o] [hieroglyph] is written exactly like [hieroglyph]; so too below 5,7.

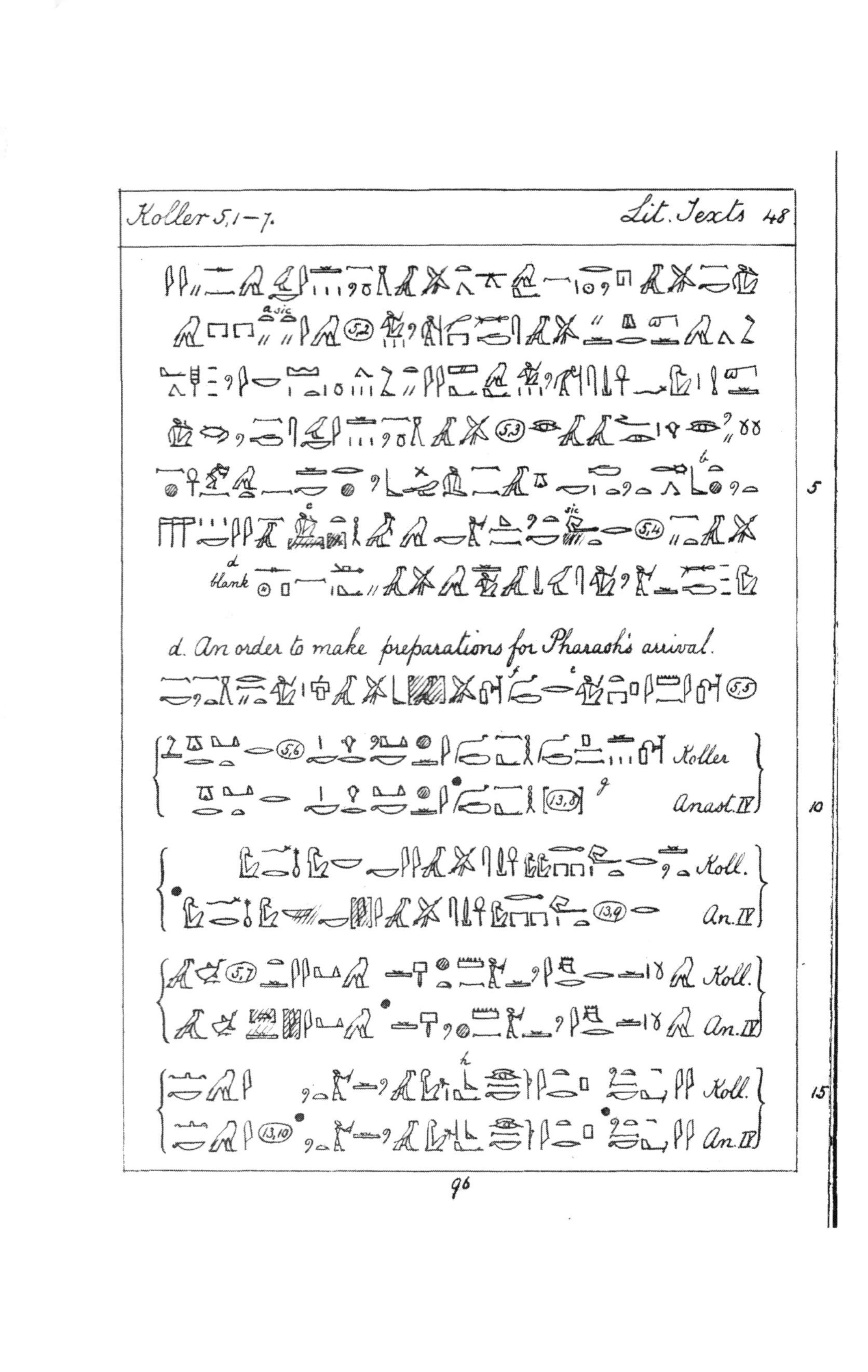

a sic

b

c

sic

(5,2) (5,3) (5,4)

5

d blank

d. An order to make preparations for Pharaoh's arrival.

f e (5,5)

(5,6) Koller

(13,8) g Anast. IV

10

Koll.

(13,9) An. IV

(5,7) Koll.

An. IV

h

Koll.

15

(13,10) An. IV

2.[a] Read [hieroglyphs].

5.[b] Evidently for [hieroglyphs], see Admonitions p. 67.

6.[c] The position of [hieroglyph] makes it certain that a sign stood beneath it.

7.[d] There is room for the sign [hieroglyph], which the scribe forgot to add.

8.[e–f] Emend [hieroglyph].

10.[g] The section begins abruptly thus without further introduction; the words [hieroglyphs] (7, 9) are probably to be understood.

15.[h] [hieroglyph] is written quite like [hieroglyph], as above 4, 8.

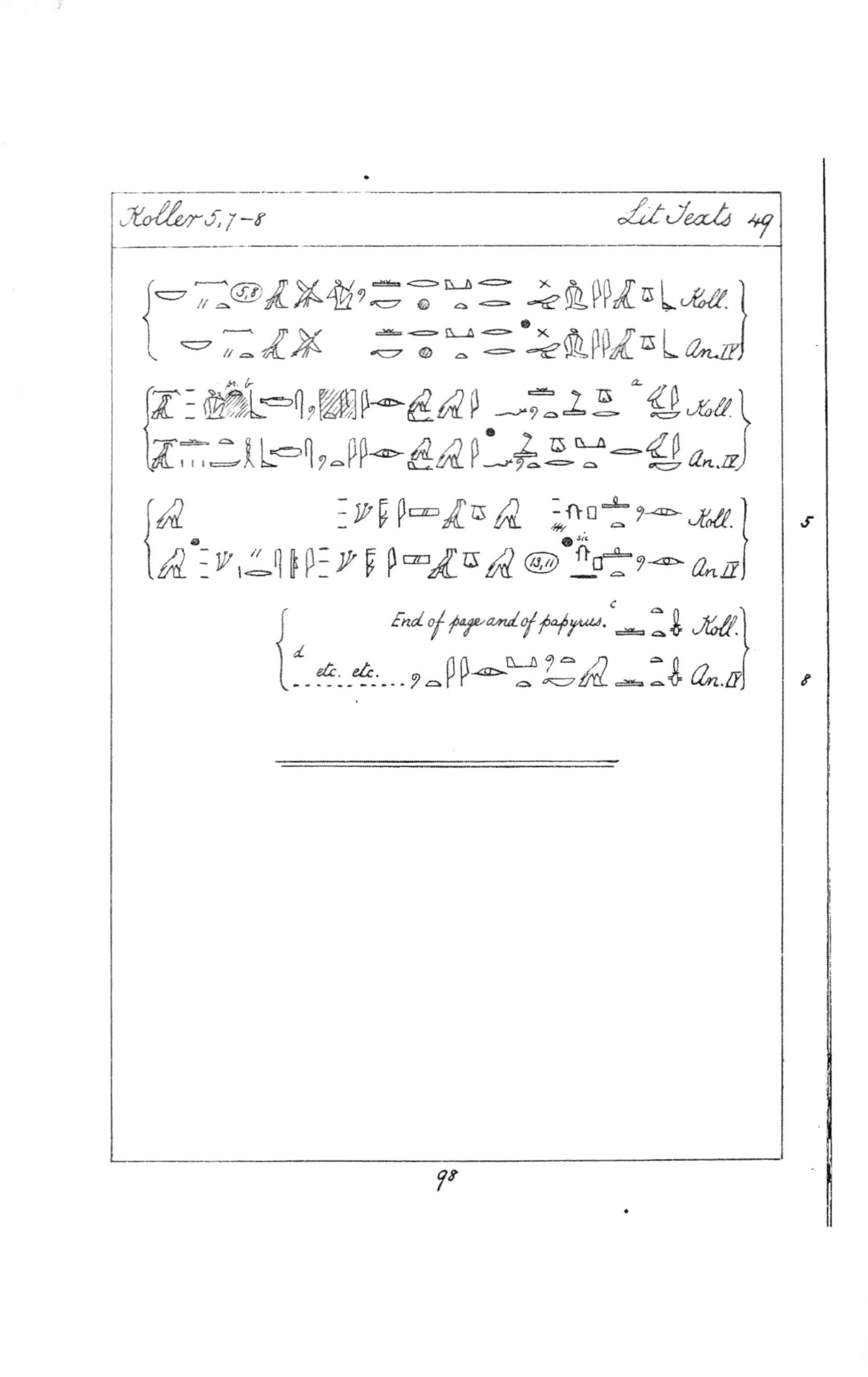

(5,8) Koll.
An. IV

pr. b a Koll.
An. IV

Koll. 5
(13,11) sic An IV

End of page and of papyrus. c Koll.
d etc. etc. .... An. IV 8

3.[a] The reading of Anast. IV is clearly superior. — [b] 𓏏 wrongly omitted.

7.[c] The text was probably continued on other pages cut off from the Pap. Koller.

8.[d] The continuation occupies the entire remainder of Anast. IV down to 17,9, where the writing suddenly stops.

Additions and Corrections.

P. 3, line 3; read [hieroglyphs]; the reading is far from certain.

P. 7, line 4; the reading [hieroglyphs] is more probable than [hieroglyphs].

P. 21, line 2; the lacuna at the end of 11,7 should be omitted; [hieroglyph] in [hieroglyphs] is nearly certain. In the note p. 21a, note 2[c], read: "corrupted out of [hieroglyphs]."

P. 32, line 18; a note on [hieroglyphs] should be added, stating that this is a wrong writing for [hieroglyphs].

P. 48, line 12, beginning; for (13,9) [hieroglyph] read (13,9) [hieroglyphs].

Milton Keynes UK
Ingram Content Group UK Ltd.
UKHW021803181223
434609UK00014B/797

9 781015 20531